Chapter 3 Mechanics 1

Chapter 4 Statistics 1

Chapter 5 Decision Mathematics 1

Specification information

AQA Mathematics

MODULE	SPECIFICATION TOPIC	CHAPTER REFERENCE	STUDIED IN CLASS	REVISED	PRACTICE QUESTIONS
Core 1	Algebra	1.1			
	Coordinate geometry	1.2			
	Differentiation	1.3			
	Integration	1.4			
Core 2	Algebra and functions	2.1			
	Sequences and series	2.2			
	Trigonometry	2.3			
	Exponentials and logarithms	2.1			
	Differentiation	2.4			
	Integration	2.5			
Mechanics 1	Vectors	3.1			
	Kinematics	3.2			
	Statics and forces	3.3			
	Momentum	3.4			
	Newton's laws of motion	3.4			
	Connected particles	3.4			
	Projectiles	3.2, 3.5			
Statistics 1	Numerical measures	4.1			
	Probability	4.2			
	Binomial distribution	4.3			
	Normal distribution	4.4			
	Estimation	4.6			
	Correlation and regression	4.5			
Decision 1	Algorithms	5.1			
	Graphs and networks	5.2			
	Spanning tree problems	5.2			
	Matchings	5.4			
	Shortest path in networks	5.2			
	Route inspection problem	5.2			
	Travelling salesperson problem	5.2			
	Linear programming	5.3			

Examination analysis

The *assessment is by written papers or a combination of written papers and coursework.*
All questions are compulsory. Core 1 + Core 2 + one of Mechanics 1, Statistics 1 and Decision 1.

Core 1	AS	No calculator	1 hr 30 min exam	33.3%
Core 2	AS	Scientific / graphic calculator	1 hr 30 min exam	33.3%
Mechanics 1	AS	Scientific / graphic calculator	1 hr 30* min exam	33.3%
Statistics 1	AS	Scientific / graphic calculator	1 hr 30* min exam	33.3%
Decision 1	AS	Scientific / graphic calculator	1 hr 30 min exam	33.3%

*1 hr 15 min with coursework.

Revise
AS

ease

AQA
Mathematics

Contents

Chapter 1 Core 1 Pure Mathematics

Chapter 2 Core 2 Pure Mathematics

Your AS/A2 Level Mathematics course

AS and A2

The AQA A Level Mathematics course is in two parts, with three separate modules in each part. Students first study the AS (Advanced Subsidiary) course. Some will then go on to study the second part of the A Level course, called A2. Advanced Subsidiary is assessed at the standard expected halfway through an A Level course: i.e., between GCSE and Advanced GCE. This means that AS and A2 courses are designed so that difficulty steadily increases:

- AS Mathematics builds from GCSE Mathematics
- A2 Mathematics builds from AS Mathematics.

How will you be tested?

Assessment units

For AS Mathematics, you will be tested by three assessment units. For the full A Level in Mathematics, you will take a further three units. AS Mathematics forms 50% of the assessment weighting for the full A Level.

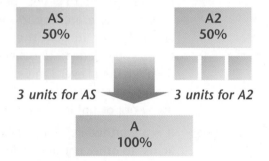

Each unit can normally be taken in either January or June. Alternatively, you can study the whole course before taking any of the unit tests. There is a lot of flexibility about when exams can be taken and the diagram below shows just some of the ways that the assessment units may be taken for AS and A Level Mathematics.

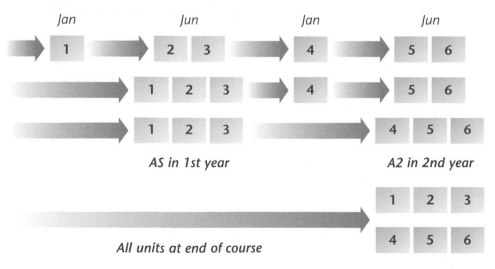

If you are disappointed with a module result, you can resit the module. There is no restriction on the number of times a module may be attempted. The best available result for each module will count towards the final grade.

Synoptic assessment

The GCE Advanced Subsidiary and Advanced Level Qualification specific Criteria state that A Level specifications must include synoptic assessment (representing at least 20% of the total A Level marks).

Synoptic assessment in mathematics addresses candidates' understanding of the connections between different elements of the subject. It involves the explicit drawing together of knowledge, understanding and skills learned in different parts of the A Level course, through using and applying methods developed at earlier stages of the course to solving problems. Making and understanding connections in this way is intrinsic to learning mathematics.

Coursework

A coursework option is available for the first mechanics and statistics components, MM1A and MS1A. The coursework carries 25% of the marks on each of these components. However, some schools choose to enter candidates for the alternative versions of these components, MM1B and MS1B, which do not include coursework.

Your school will inform you about any coursework requirements.

Key skills

It is important that you develop your key skills throughout your AS and A2 courses. These are important skills that you need whatever you do beyond AS and A Levels. To gain the key skills qualification, which is equivalent to an AS Level, you will need to collect evidence together in a 'portfolio' to show that you have attained Level 3 in Communication, Application of number and Information technology. You will also need to take a formal test in each key skill. You will have many opportunities during AS Mathematics to develop your key skills.

It is a worthwhile qualification, as it demonstrates your ability to put your ideas across to other people, collect data and use up-to-date technology in your work.

What skills will I need?

For AS Mathematics, you will be tested by assessment objectives: these are the skills and abilities that you should have acquired by studying the course. The assessment objectives for AS Mathematics are shown below.

Candidates should be able to:

- recall, select and use their knowledge of mathematical facts, concepts and techniques in a variety of contexts
- construct rigorous mathematical arguments and proofs through use of precise statements, logical deduction and inference and by the manipulation of mathematical expressions, including the construction of extended arguments for handling substantial problems presented in unstructured form

- recall, select and use their knowledge of standard mathematical models to represent situations in the real world; recognise and understand given representations involving standard models; present and interpret results from such models in terms of the original situation, including discussion of the assumptions made and refinement of such models
- comprehend translations of common realistic contexts into Mathematics; use the results of calculations to make predictions, or comment on the context; and, where appropriate, read critically and comprehend longer mathematical arguments or examples of applications
- use contemporary calculator technology and other permitted resources (such as formulae booklets or statistical tables*) accurately and efficiently; understand when not to use such technology, and its limitations; give answers to appropriate accuracy.

*A copy of the formulae booklet and statistical tables can be found on the A Level mathematics section of the AQA website, www.aqa.org.uk

Progression and prior learning

Mathematics is, inherently, a sequential subject. There is a progression of material through all levels at which the subject is studied. The criteria therefore build on the knowledge, understanding and skills established at GCSE.

Thus, candidates embarking on AS/A2 Level study in Mathematics subjects are expected to have covered all the material in the Higher Tier. This material is regarded as assumed background knowledge. However, it may be assessed within questions focused on other material from the specification.

Exam technique

Examiners use certain words in their instructions to let you know what they are expecting in your answer. Make sure that you know what they mean so that you can give the right response.

Write down, state

You can write your answer without having to show how it was obtained. There is nothing to prevent you doing some working if it helps you, but if you are doing a lot then you might have missed the point.

Calculate, find, determine, show, solve

Make sure that you show enough working to justify the final answer or conclusion. Marks will be available for showing a correct method.

Deduce, hence

This means that you are expected to use the given result to establish something new. You must show all of the steps in your working.

Draw

This is used to tell you to plot an accurate graph using graph paper. Take note of any instructions about the scale that must be used. You may need to read values from your graph.

Sketch

If the instruction is to sketch a graph then you don't need to plot the points but you will be expected to show its general shape and its relationship with the axes. Indicate the positions of any turning points and take particular care with any asymptotes.

Find the exact value

This instruction is usually given when the final answer involves an irrational value such as a logarithm, π or a surd. You will need to demonstrate that you can manipulate these quantities so don't just key everything into your calculator or you will lose marks.

If a question requires the final answer to be given to a specific level of accuracy then make sure that you do this or you might needlessly lose marks.

Some dos and don'ts

Dos

Do read the question

- Make sure that you are clear about what you are expected to do. Look for some structure in the question that may help you take the right approach.
- Read the question *again* after you have answered it as a quick check that your answer is in the expected form.

Do use diagrams

- In some questions, particularly in mechanics, a clearly labelled diagram is essential. Use a diagram whenever it may help you understand or represent the problem that you are trying to solve.

Do take care with notation

- Write clearly and use the notation accurately. Use brackets when they are required.
- Even if your final answer is wrong, you may earn some marks for a correct expression in your working.

Do avoid silly answers

- Check that your final answer is sensible within the context of the question.

Do make good use of time

- Choose the order in which you answer the questions carefully. Do the ones that are easiest for you first.
- Set yourself a time limit for a question depending on the number of marks available.
- Be prepared to leave a difficult part of a question and return to it later if there is time.
- Towards the end of the exam make sure that you pick up all of the easy marks in any questions that you haven't got time to answer fully.

Don'ts

Don't work with rounded values

- There may be several stages in a solution that produce numerical values. Rounding errors from earlier stages may distort your final answer. One way to avoid this is to make use of your calculator memories to store values that you will need again.

Don't cross out work that may be partly correct

- It's tempting to cross out something that hasn't worked out as it should. Avoid this unless you have time to replace it with something better.

Don't write out the question

- This wastes time. The marks are for your solution!

What grade do you want?

Everyone should be able to improve their grades but you will only manage this with a lot of hard work and determination. The details given below describe a level of performance typical of candidates achieving grades A, C or E. You should find it useful to read and compare the expectations for the different levels and to give some thought to the areas where you need to improve most.

Grade A candidates

- Recall or recognise almost all the mathematical facts, concepts and techniques that are needed, and select appropriate ones to use in a variety of contexts.
- Manipulate mathematical expressions and use graphs, sketches and diagrams, all with high accuracy and skill.
- Use mathematical language correctly and proceed logically and rigorously through extended arguments or proofs.
- When confronted with unstructured problems they can often devise and implement an effective solution strategy.
- If errors are made in their calculations or logic, these are sometimes noticed and corrected.
- Recall or recognise almost all the standard models that are needed, and select appropriate ones to represent a wide variety of situations in the real world.
- Correctly refer results from calculations using the model to the original situation; they give sensible interpretations of their results in the context of the original realistic situation.
- Make intelligent comments on the modelling assumptions and possible refinements to the model.
- Comprehend or understand the meaning of almost all translations into mathematics of common realistic contexts.
- Correctly refer the results of calculations back to given context and usually make sensible comments or predictions.
- Can distil the essential mathematical information from extended pieces of prose having mathematical content.
- Comment meaningfully on the mathematical information.
- Make appropriate and efficient use of contemporary calculator technology and other permitted resources, and are aware of any limitations to their use.
- Present results to an appropriate degree of accuracy.

Grade C candidates

- Recall or recognise most of the mathematical facts, concepts and techniques that are needed, and usually select appropriate ones to use in a variety of contexts.
- Manipulate mathematical expressions and use graphs, sketches and diagrams, all with a reasonable level of accuracy and skill.
- Use mathematical language with some skill and sometimes proceed logically through extended arguments or proofs.
- When confronted with unstructured problems they sometimes devise and implement an effective and efficient solution strategy.
- Occasionally notice and correct errors in their calculations.
- Recall or recognise most of the standard models that are needed and usually select appropriate ones to represent a variety of situations in the real world.

- Often correctly refer results from calculations using the model to the original situation; they sometimes give sensible interpretations of their results in context of the original realistic situation.
- Sometimes make intelligent comments on the modelling assumptions and possible refinements to the model.
- Comprehend or understand the meaning of most translations into mathematics of common realistic contexts.
- Often correctly refer the results of calculations back to the given context and sometimes make sensible comments or predictions.
- Distil much of the essential mathematical information from extended pieces of prose having mathematical content.
- Give some useful comments on this mathematical information.
- Usually make appropriate and effective use of contemporary calculator technology and other permitted resources, and are sometimes aware of any limitations to their use.
- Usually present results to an appropriate degree of accuracy.

Grade E candidates

- Recall or recognise some of the mathematical facts, concepts and techniques that are needed, and sometimes select appropriate ones to use in some contexts.
- Manipulate mathematical expressions and use graphs, sketches and diagrams, all with some accuracy and skill.
- Sometimes use mathematical language correctly and occasionally proceed logically through extended arguments or proofs.
- Recall or recognise some of the standard models that are needed and sometimes select appropriate ones to represent a variety of situations in the real world.
- Sometimes correctly refer results from calculations using the model to the original situation; they try to interpret their results in the context of the original realistic situation.
- Sometimes comprehend or understand the meaning of translations into mathematics of common realistic contexts.
- Sometimes correctly refer the results of calculations back to the given context and attempt to give comments or predictions.
- Distil some of the essential mathematical information from extended pieces of prose having mathematical content; they attempt to comment on this mathematical information.
- Candidates often make appropriate and efficient use of contemporary calculator technology and other permitted resources.
- Often present results to an appropriate degree of accuracy.

The table below shows how your uniform standardised mark is translated.

average	80%	70%	60%	50%	40%
grade	A	B	C	D	E

To achieve an A* grade in Mathematics, you need to achieve a grade A overall (an average of 80 or more on uniform mark scale) for the <u>whole</u> A Level qualification and an average of 90 or more on the uniform mark scale in Core 3 and Core 4. It is awarded for A Level qualification only and not for the AS qualification or individual units.

Four steps to successful revision

Step 1: Understand

- Study the topic to be learned slowly. Make sure you understand the logic or important concepts.
- Mark up the text if necessary – underline, highlight and make notes.
- Re-read each paragraph slowly.

GO TO STEP 2

Step 2: Summarise

- Now make your own revision note summary:
  What is the main idea, theme or concept to be learned?
  What are the main points? How does the logic develop?
  Ask questions: Why? How? What next?
- Use bullet points, mind maps, patterned notes.
- Link ideas with mnemonics, mind maps, crazy stories.
- Note the title and date of the revision notes
  (e.g. Mathematics: Trigonometry, 3rd March).
- Organise your notes carefully and keep them in a file.

This is now in **short-term memory**. You will forget 80% of it if you do not go to Step 3. *GO TO STEP 3*, but first take a 10 minute break.

Step 3: Memorise

- Take 25 minute learning 'bites' with 5 minute breaks.
- After each 5 minute break test yourself:
  Cover the original revision note summary
  Write down the main points
  Speak out loud (record yourself)
  Tell someone else
  Repeat many times.

The material is well on its way to **long-term memory**. You will forget 40% if you do not do step 4. *GO TO STEP 4*

Step 4: Track/Review

- Create a Revision Diary (one A4 page per day).
- Make a revision plan for the topic, e.g. 1 day later, 1 week later, 1 month later.
- Record your revision in your Revision Diary, e.g.
  Mathematics: Trigonometry, 3rd March 25 minutes
  Mathematics: Trigonometry, 5th March 15 minutes
  Mathematics: Trigonometry, 3rd April 15 minutes
  ... and then at monthly intervals.

Chapter 1

Core 1 Pure Mathematics

The following topics are covered in this chapter:

- *Algebra and functions*
- *Coordinate geometry*
- *Differentiation*
- *Integration*

Note: Calculators are *not* allowed in this module

1.1 Algebra and functions

After studying this section you should be able to:

- work with surds
- use function notation
- solve quadratic equations and sketch the graphs of quadratic functions
- understand the definition of a polynomial
- carry out algebraic division
- use the remainder and factor theorems
- recognise and sketch graphs of a range of functions
- understand how a translation affects the equation of a curve
- solve simultaneous equations – one linear and one quadratic
- solve linear and quadratic inequalities

LEARNING SUMMARY

Surds

A **surd** is the root of a whole number that has an **irrational** value.

Some examples are $\sqrt{2}$, $\sqrt{3}$ and $\sqrt{10}$.

You can often simplify a surd using the fact that $\sqrt{ab} = \sqrt{a} \times \sqrt{b}$ and choosing a or b to be a square number.

For example $\sqrt{12} = \sqrt{4 \times 3} = \sqrt{4} \times \sqrt{3} = 2\sqrt{3}$ and $\sqrt{18} + 2\sqrt{32} = 3\sqrt{2} + 2 \times 4\sqrt{2} = 11\sqrt{2}$.

> An irrational number continues for ever after the decimal point without a repeating pattern.

To simplify $\dfrac{3}{\sqrt{5}}$, multiply the numerator and the denominator by $\sqrt{5}$ to get

$\dfrac{3}{\sqrt{5}} \times \dfrac{\sqrt{5}}{\sqrt{5}} = \dfrac{3\sqrt{5}}{5}$. This is called **rationalising the denominator**.

Functions

A **function** may be thought of as a rule which takes each member x of a set and assigns, or **maps**, it to some value y known as its **image**.

> x maps to y
> y is the image of x.

$$x \longrightarrow \boxed{\text{Function}} \longrightarrow y$$

A letter such as f, g or h is often used to stand for a function. The function which squares a number and adds on 5, for example, can be written as $f(x) = x^2 + 5$. The same notation may also be used to show how a function affects particular values.

> $f(x)$ is read as 'f of x'.

For this function, $f(4) = 4^2 + 5 = 21$, $f(-10) = (-10)^2 + 5 = 105$ and so on.

An alternative notation for the same function is $f: x \mapsto x^2 + 5$.

Quadratics

In algebra, any expression of the form $ax^2 + bx + c$ where $a \neq 0$ is called a **quadratic expression**.

You need to be able to expand brackets, **for example**

$$(2x + 3)(x - 4) = 2x^2 - 8x + 3x - 12$$
$$= 2x^2 - 5x - 12$$

It is useful to remember these special results:

$$(x + a)^2 = x^2 + 2ax + a^2 \qquad (x - a)^2 = x^2 - 2ax + a^2 \qquad (x + a)(x - a) = x^2 - a^2$$

Some examples are:

$$(x + 3)^2 = x^2 + 6x + 9 \qquad\qquad (x - 5)^2 = x^2 - 10x + 25$$
$$(x + 7)(x - 7) = x^2 - 49 \qquad\qquad (x + \sqrt{3})(x - \sqrt{3}) = x^2 - 3$$

To solve problems in algebra you need to develop your skills so that you can recognise how to apply the basic results.

> In this fraction, the denominator $2 - \sqrt{3}$ is irrational.

> This is an example of rationalising the denominator.

For example, the fraction $\dfrac{1}{2 - \sqrt{3}}$ may be simplified by multiplying the numerator and denominator by $2 + \sqrt{3}$ to make an equivalent fraction in which the denominator is rational.

> $(2 - \sqrt{3})(2 + \sqrt{3})$
> $= 2^2 - (\sqrt{3})^2$
> $= 4 - 3$.

$$\frac{1}{2 - \sqrt{3}} = \frac{1}{2 - \sqrt{3}} \times \frac{2 + \sqrt{3}}{2 + \sqrt{3}}$$

$$= \frac{2 + \sqrt{3}}{4 - 3} \qquad \text{(The denominator is now rational)}$$

$$= 2 + \sqrt{3}$$

Quadratic equations

Equations of the form $ax^2 + bx + c = 0$ (where $a \neq 0$) are **quadratic equations**.

Some quadratic equations can be solved by **factorising** the quadratic expression.

Example Solve $2x^2 - x - 3 = 0$.

> Factorise.

$$(2x - 3)(x + 1) = 0$$

Either $2x - 3 = 0$ or $x + 1 = 0$

so $x = \frac{3}{2}$ or $x = -1$.

Some equations can be converted into a quadratic equation by substitution.

Example Solve $x^{\frac{2}{3}} + x^{\frac{1}{3}} - 6 = 0$

Substituting $y = x^{\frac{1}{3}}$ gives

$$y^2 + y - 6 = 0$$

$$(y + 3)(y - 2) = 0$$

Either $y + 3 = 0$ or $y - 2 = 0$

so $y = -3$ or $y = 2$.

Re-writing in terms of x gives $x^{\frac{1}{3}} = -3$ or $x^{\frac{1}{3}} = 2$,

so $x = -27$ or $x = 8$.

If the quadratic will not factorise then you can try **completing the square**.

Example Solve $x^2 - 6x + 1 = 0$

$x^2 - 6x = (x - 3)^2 - 9$ so
$x^2 - 6x + 1 = (x - 3)^2 - 8$.

$(x - 3)^2 - 8 = 0$ (Now x only appears once in the equation)

$(x - 3)^2 = 8$

$x - 3 = \pm\sqrt{8}$

$\sqrt{8} = \sqrt{(4 \times 2)} = 2\sqrt{2}$.

$x = 3 \pm 2\sqrt{2}$.

(The solutions are $3 + 2\sqrt{2}$ and $3 - 2\sqrt{2}$)

The method shown for completing the square can be adapted for the general form of a quadratic. This gives the **quadratic formula**.

$ax^2 + bx + c = 0$

Multiply both sides by $4a$.

$4a^2x^2 + 4abx + 4ac = 0$ Note: $4a^2x^2 + 4abx + b^2 = (2ax + b)^2$.

Add b^2 to complete the square and subtract it again to keep things the same.

$4a^2x^2 + 4abx + b^2 + 4ac - b^2 = 0$
$(2ax + b)^2 + 4ac - b^2 = 0$ Now x only appears once in the equation.
$(2ax + b)^2 = b^2 - 4ac$

Find the square root of both sides.

$2ax + b = \pm\sqrt{b^2 - 4ac}$

Rearrange to find x.

$x = \dfrac{-b \pm \sqrt{b^2 - 4ac}}{2a}$ You must learn this formula.

Example Solve $3x^2 - 4x - 2 = 0$ giving your answers in surd form.

Comparing this equation with the general form gives $a = 3$, $b = -4$ and $c = -2$. Substitute this information into the formula:

In Core 1 you do not have access to a calculator so you will have to leave the answer in surd form.

$x = \dfrac{4 \pm \sqrt{(-4)^2 - 4(3)(-2)}}{6} = \dfrac{4 \pm \sqrt{40}}{6} = \dfrac{4 \pm 2\sqrt{10}}{6} = \dfrac{2 \pm \sqrt{10}}{3}$

so $x = \dfrac{2 + \sqrt{10}}{3}$ or $x = \dfrac{2 - \sqrt{10}}{3}$.

Discriminant

In the formula, the value of $b^2 - 4ac$ is called the **discriminant**. This value can be used to give information about the solutions without having to solve the equation.

> In the previous example, $b^2 - 4ac = 40 > 0$ and two real roots were found.

$b^2 - 4ac > 0$ *two* distinct real solutions (the solutions are often called **roots**)

$b^2 - 4ac = 0$ *one* real solution (often thought of as a **repeated root**)

$b^2 - 4ac < 0$ *no* real solutions.

The solutions of a quadratic equation correspond to where the graph of the quadratic function crosses the x-axis. There are three possible situations depending on the value of the discriminant.

> The diagrams correspond to the situation where $a > 0$. The same principle applies when $a < 0$ but the diagrams appear the other way up.

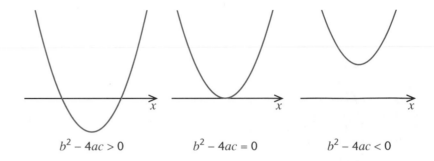

$b^2 - 4ac > 0$ $b^2 - 4ac = 0$ $b^2 - 4ac < 0$

Example   The equation $5x^2 + 3x + p = 0$ has a repeated root. Find the value of p.

In this case, $a = 5$, $b = 3$ and $c = p$.

For a repeated root $b^2 - 4ac = 0$ so $9 - 20p = 0$, giving $p = 0.45$

Quadratic graphs

The graph of $y = ax^2 + bx + c$ is a parabola.

$a > 0$ $a < 0$

The methods used for solving quadratic equations can also be used to give information about the graphs.

Example   Sketch the graph of $y = x^2 - x - 6$.
           Find the coordinates of the lowest point on the curve.

The curve will cross the x-axis when $y = 0$. You can find these points by solving the equation $x^2 - x - 6 = 0$.

> The curve will cross the y-axis when $x = 0$, giving $y = -6$.

$x^2 - x - 6 = 0 \implies (x + 2)(x - 3) = 0$

$\implies x = -2$ or $x = 3$

The curve is symmetrical so the lowest point occurs mid-way between -2 and 3 and this is given by $(-2 + 3) \div 2 = 0.5$

When $x = 0.5$, $y = 0.5^2 - 0.5 - 6 = -6.25$

The lowest point on the curve is $(0.5, -6.25)$.

> The vertex of the curve occurs at its maximum or minimum point.

Completing the square gives information about the **vertex** of the curve even if the equation will not factorise.

Example Find the coordinates of the vertex of the curve $y = x^2 + 2x + 3$.

You need to recognise that $x^2 + 2x + 1 = (x + 1)^2$,

then, completing the square, $x^2 + 2x + 3 = x^2 + 2x + 1 + 2 = (x + 1)^2 + 2$.

The equation of the curve can now be written as $y = (x + 1)^2 + 2$.

$(x + 1)^2$ cannot be negative so its minimum value is zero, when $x = -1$.

This means that the minimum value of y is 2 and this occurs when $x = -1$.

The vertex of the curve is at $(-1, 2)$.

Polynomials

A quadratic is one example of a **polynomial**. In general, a polynomial takes the form:

$$a_n x^n + a_{n-1} x^{n-1} + a_{n-2} x^{n-2} + \ldots + a_0,$$

> This is much simpler than it looks.

where $a_n, a_{n-1}, \ldots a_0$ are constants and n is a positive whole number.

For example, $x^4 - 2x^3$ is a polynomial. In this case $a_4 = 1$, $a_3 = -2$ and a_2, a_1, a_0 are all zero.

The **degree** of a polynomial is the highest power of x that it includes, so the degree of $x^4 - 2x^3$ is 4. A quadratic is a polynomial of degree 2, a cubic is a polynomial of degree 3 and so on.

Algebraic division

If you work out $27 \div 4$ for example, then you obtain 6 as the **quotient** and 3 as the **remainder**. You can write $27 = 4 \times 6 + 3$ to show the connection between these values.
The same thing applies in algebra when one polynomial is divided by another.

Example Find the quotient and remainder when $x^2 + 7x - 5$ is divided by $x - 4$.

Using $ax + b$ for the quotient and c for the remainder gives:

$$x^2 + 7x - 5 \equiv (x - 4)(ax + b) + c.$$
$$\equiv ax^2 + (b - 4a)x + c - 4b.$$

Equating coefficients of x^2 gives: $a = 1$.

Equating coefficients of x gives: $b - 4a = 7$

$$\Rightarrow b - 4 = 7 \qquad \text{(using the result } a = 1 \text{ from above)}$$

$$\Rightarrow b = 11.$$

Equating the constant terms gives: $c - 4b = -5$

$$\Rightarrow c - 44 = -5 \quad \text{(using the result } b = 11 \text{ from above)}$$

$$\Rightarrow c = 39.$$

Substituting for a, b and c in [1] gives $x^2 + 7x - 5 \equiv (x - 4)(x + 11) + 39$.

The remainder theorem

In the previous example, the quotient and remainder of $(x^2 + 7x - 5) \div (x - 4)$ were found by using $x^2 + 7x - 5 \equiv (x - 4)(ax + b) + c$ and equating coefficients.

When $x = 4$, $(x - 4) = 0$ so $(x - 4)(ax + b) = 0$ and the RHS simplifies to c.

Another way to use this identity is to substitute particular values for x. Notice that when $x = 4$, the RHS simplifies to c and so the remainder is easily found by substituting $x = 4$ on the LHS. This gives $c = 4^2 + 7 \times 4 - 5 = 39$ as before.

In its generalised form this result is known as the **remainder theorem** which states:

> When a polynomial $f(x)$ is divided by $(x - a)$ the remainder is $f(a)$.   **KEY POINT**

Example Find the remainder when the polynomial $f(x) = x^3 + x - 5$ is divided by:

(a) $(x - 3)$ (b) $(x + 2)$ (c) $(2x - 1)$.

Choose x so that $x - 3 = 0$.

(a) The remainder is $f(3) = 3^3 + 3 - 5 = 25$.

(b) The remainder is $f(-2) = (-2)^3 - 2 - 5 = -15$.

Choose x so that $2x - 1 = 0$.

(c) The remainder is $f(0.5) = 0.5^3 + 0.5 - 5 = -4.375$

The factor theorem

A special case of the remainder theorem occurs when the remainder is 0. This result is known as the **factor theorem** which states:

> If $f(x)$ is a polynomial and $f(a) = 0$ then $(x - a)$ is a factor of $f(x)$.   **KEY POINT**

Example Show that $(x + 3)$ is a factor of $x^3 + 5x^2 + 5x - 3$.

$$\text{Taking } f(x) = x^3 + 5x^2 + 5x - 3$$

$$f(-3) = (-3)^3 + 5(-3)^2 + 5(-3) - 3$$

$$= -27 + 45 - 15 - 3 = 0.$$

By the factor theorem $(x + 3)$ is a factor of $x^3 + 5x^2 + 5x - 3$.

Example A polynomial is given by $f(x) = 2x^3 + 13x^2 + 13x - 10$.

      (a) Find the value of $f(2)$ and $f(-2)$.

      (b) State one of the factors of $f(x)$.

      (c) Factorise $f(x)$ completely.

(a) $f(2) = 2(2^3) + 13(2^2) + 13(2) - 10 = 84$

      $f(-2) = 2(-2)^3 + 13(-2)^2 + 13(-2) - 10 = 0$.

(b) Since $f(-2) = 0$, $(x + 2)$ is a factor of $f(x)$ by the factor theorem.

> $f(x)$ is a cubic so the linear factor $(x + 2)$ must be combined with a quadratic.

(c) $2x^3 + 13x^2 + 13x - 10 \equiv (x + 2)(ax^2 + bx + c)$.

This is an **identity** and so the values of a, b and c may be found by comparing coefficients.

Comparing the x^3 terms gives:	$a = 2$.
Comparing the x^2 terms gives:	$13 = 2a + b \Rightarrow b = 9$.
Comparing the constant terms gives:	$c = -5$.
It follows that:	$f(x) = (x + 2)(2x^2 + 9x - 5)$.

Factorising the quadratic part in the usual way gives $f(x) = (x + 2)(2x - 1)(x + 5)$.

The graph of a cubic

The graph of a cubic function $y = ax^3 + bx^2 + cx + d$ can take a number of forms.

$a > 0$ $a < 0$

> **KEY POINT**
>
> The graph of a cubic function that can be factorised as $y = (x - p)(x - q)(x - r)$ will cross the x-axis at p, q and r. If any two of p, q and r are the same then the x-axis will be a tangent to the curve at that point.

For example, the graph of $y = (x + 2)(x - 3)^2$ looks like this:

> This shows a repeated root of $(x + 2)(x - 3)^2 = 0$ at $x = 3$.

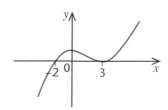

The graph of $y = x^n$ where n is an integer has:
rotational symmetry about the **origin** when n is **odd**
reflective symmetry in the y-axis when n is **even**.

For higher powers of x the graphs are flatter between −1 and 1 and steeper elsewhere.

The graphs of $y = x^3$, $y = x^5$, $y = x^7$ ... look something like this:

The graphs of $y = x^2$, $y = x^4$, $y = x^6$ ... look something like this:

A straight line that a graph approaches ever more closely without actually touching it is called an **asymptote**.

As the powers of x become more negative, the graphs become steeper between −1 and 1 and flatter elsewhere.

The graphs of $y = x^{-1}$, $y = x^{-3}$, $y = x^{-5}$ ... look something like this:

The graphs of $y = x^{-2}$, $y = x^{-4}$, $y = x^{-6}$ ... look something like this:

Both the x and y-axes are asymptotes for these graphs.

The x-axis and the positive y-axis are asymptotes for these graphs.

The graphs of $y = k\sqrt{x}$, $k > 0$, look something like this. These graphs do not have any asymptotes.

Translations, graphs and equations

For any function, f, the graph of $y = f(x - a) + b$ is obtained from the graph of $y = f(x)$ by the translation $\begin{bmatrix} a \\ b \end{bmatrix}$.

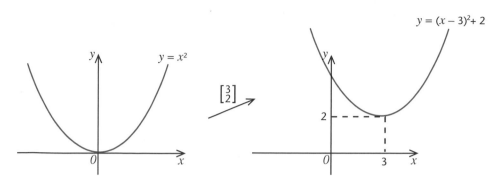

Simultaneous equations

The substitution method of solving pairs of linear simultaneous equations can also be applied when one of the equations is a quadratic.

Example   Find the coordinates of the points where the line $y = x + 1$ intersects the circle $x^2 + y^2 = 5$.

$$y = x + 1 \quad (1)$$
$$x^2 + y^2 = 5 \quad (2)$$

Substitute for y from (1) into (2): $x^2 + (x + 1)^2 = 5$.

Now expand the brackets: $x^2 + x^2 + 2x + 1 = 5$.

Arrange in the form $ax^2 + bx + c = 0$: $2x^2 + 2x - 4 = 0$.

Divide by 2. $x^2 + x - 2 = 0$

Solve the equation: $(x - 1)(x + 2) = 0$
Either $x = 1$ or $x = -2$

Substitute into (1) to find the y values: When $x = 1$, $y = 2$
When $x = -2$, $y = -1$

The coordinates of the points of intersection are (1, 2) and (−2, −1).

Geometrical interpretation of algebraic solutions

In the above example, when $y = x + 1$ and $x^2 + y^2 = 5$ are solved simultaneously, the resulting quadratic equation in x has two distinct solutions. This gives the two points of intersection of the line and the curve.

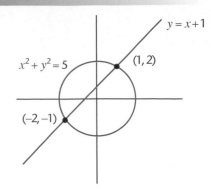

When you solve the equations of a line and a curve simultaneously and form a quadratic equation, $ax^2 + bx + c = 0$, the discriminant, $b^2 - 4ac$, gives information about the number of points of intersection.

If $b^2 - 4ac > 0$ the line and curve intersect in two distinct points.

If $b^2 - 4ac = 0$ the line is a tangent to the curve.

If $b^2 - 4ac < 0$ the line and the curve do not intersect.

Inequalities

Linear inequalities can be solved by rearrangement in much the same way as linear equations. However, care must be taken to reverse the direction of the inequality when multiplying or dividing by a negative.

Example   Solve the inequality $8 - 3x > 23$.

Subtract 8 from both sides: $-3x > 15$
Divide both sides by -3: $x < -5$.

An inequality which has x on both sides is treated like the corresponding equation.

Example   Solve the inequality $5x - 3 > 3x - 10$.

Subtract $3x$ from both sides: $2x - 3 > -10$
Add 3 to both sides: $2x > -7$
Divide both sides by 2: $x > -3.5$

Quadratic inequalities are solved in a similar way to quadratic equations but a sketch graph is often helpful at the final stage.

Example   Solve the inequality $x^2 - 3x + 2 < 0$.

Factorise the quadratic expression: $(x - 1)(x - 2) < 0$.

Sketch the graph of $y = (x - 1)(x - 2)$:

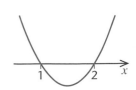

$y = 0$ when $x = 1$ or when $x = 2$. The graph must cross the x-axis at these points.

$(x - 1)(x - 2) < 0$ when the curve is below the x-axis.

The graph shows that $(x - 1)(x - 2) < 0$ for x values between 1 and 2.

It follows that $x^2 - 3x + 2 < 0$ when $1 < x < 2$.

If the quadratic expression cannot be factorised then the formula may be used to find the points of intersection of the curve with the x-axis.

Example Solve the inequality $x^2 + 2x - 5 > 0$.

$\sqrt{24} = \sqrt{4 \times 6} = 2\sqrt{6}$

Use the formula to solve $x^2 + 2x - 5 = 0$. $x = \dfrac{-2 \pm \sqrt{24}}{2} = \dfrac{-2 \pm 2\sqrt{6}}{2}$

Simplify the result. $x = -1 \pm \sqrt{6}$

$x^2 + 2x - 5 > 0$ when the curve is above the x-axis.

Sketch the graph of $y = x^2 + 2x - 5$:

Write the solution as two separate inequalities.

From the sketch, $x^2 + 2x - 5 > 0$ when $x < -1 - \sqrt{6}$ or when $x > -1 + \sqrt{6}$.

Progress check

1   Simplify these surd expressions:

 (a) $\sqrt{72}$ (b) $(1 + \sqrt{3})(1 - \sqrt{3})$ (c) $5\sqrt{12} - 6\sqrt{3}$.

2   (a) Solve $2x^2 + x - 21 = 0$ by factorising.
 (b) Solve $x^2 - 6x + 7 = 0$ by completing the square.
     Give the roots in surd form.
 (c) Solve $3x^2 + 2x - 2 = 0$ using the quadratic formula.
     Give the roots in surd form.

3   Express $x^2 - 8x + 17$ in the form $(x - a)^2 + b$ and hence write down its minimum value.

4   Sketch the graph of $y = (x + 3)(x - 2)^2$.

5   Solve these equations simultaneously.
    $x + 2y = 6$
    $4y^2 - 5x^2 = 36$.

6   Solve these inequalities.

 (a) $3x - 7 > 5x + 11$ (b) $2x^2 - 3x - 5 \leqslant 0$.

7   Find the remainder when $x^3 + 2x^2 - 5x + 1$ is divided by $(x - 2)$.

7   7

6   (a) $x > -9$ (b) $-1 \leqslant x \leqslant 2.5$

5   $x = 0, y = 3$ or $x = -3, y = 4.5$

3   $(x - 4)^2 + 1$, min value = 1

2   (a) $3, -3.5$ (b) $3 \pm \sqrt{2}$ (c) $\dfrac{-1 \pm \sqrt{7}}{3}$

1   (a) $6\sqrt{2}$ (b) -2 (c) $4\sqrt{3}$

1.2 Coordinate geometry

After studying this section you should be able to:

- *find the gradient of a line joining two points*
- *recognise equations of straight lines in various forms*
- *construct the equation of a straight line parallel to a given line and passing through a given point*
- *construct the equation of a straight line perpendicular to a given line and passing through a given point*
- *construct the equation of a straight line passing through two given points*
- *find the coordinates of the mid-point of two given points*
- *find the distance between any two given points*
- *use the equation of a circle and circle properties*

Gradient of a line

> **KEY POINT**
>
> The gradient of a line joining the points (x_1, y_1) and (x_2, y_2) is given by the formula $m = \dfrac{y_2 - y_1}{x_2 - x_1}$

For example, the gradient of the line joining $(2, -5)$ and $(-1, 4)$ is
$$\frac{4 - (-5)}{-1 - 2} = \frac{9}{-3} = -3.$$

Straight lines

The general equation of a straight line is $ax + by + c = 0$. The equation of any straight line can be written in this form.

For example, the line $x + y = 5$ corresponds to $a = 1$, $b = 1$ and $c = -5$.

> The gradient of a vertical line is undefined.

Straight lines, apart from those parallel to the y-axis, can be written in the form $y = mx + c$. This is known as **gradient–intercept** form because the gradient (m) and the y-intercept (c) are clearly identified in the equation. This makes it easy to construct an equation when the gradient and intercept are known.

For example, the line with gradient 4 crossing the y-axis at -5 has equation $y = 4x - 5$.

> **KEY POINT**
>
> Straight lines that are **parallel** must have the **same gradient**.

Example   Find the equation of the straight line parallel to $y = 3x - 5$ and passing through the point $(4, 2)$.

> This equation must be satisfied at the point where $x = 4$ and $y = 2$.

The line must have gradient 3 and so it can be written in the form $y = 3x + c$.
Substituting $x = 4$ and $y = 2$ gives $2 = 3 \times 4 + c \Rightarrow c = -10$.
The required equation is $y = 3x - 10$.

> **KEY POINT**
>
> The equation of a straight line with gradient m and passing through the point (x_1, y_1) can be written as $y - y_1 = m(x - x_1)$.

Using this in the example above gives $y - 2 = 3(x - 4)$. The equation is acceptable in this form but it can be rearranged to give $y = 3x - 10$ as before.

Example Find the equation of the straight line passing through the points $(-1, 5)$ and $(3, -2)$.

One approach is to use the form $y = mx + c$ to produce a pair of simultaneous equations:

Substituting $x = -1$ and $y = 5$ gives $5 = -m + c$ (1)
Substituting $x = 3$ and $y = -2$ gives $-2 = 3m + c$ (2)

(2) – (1) gives: $-7 = 4m \Rightarrow m = \dfrac{-7}{4}$.

Substituting for m in (1) gives: $5 = \dfrac{7}{4} + c \Rightarrow c = \dfrac{13}{4}$.

The equation of the line is $y = \dfrac{-7}{4}x + \dfrac{13}{4}$. This is the same as $4y + 7x = 13$.

An alternative approach is to find the gradient directly and then use the form $y - y_1 = m(x - x_1)$.

Taking $(-1, 5)$ as (x_1, y_1) and $(3, -2)$ as (x_2, y_2),
$$m = \frac{-2 - 5}{3 - (-1)} = \frac{-7}{4} = -\frac{7}{4}.$$

The equation of the line is $y - 5 = -\dfrac{7}{4}(x + 1)$.

To show that this is the same as the previous result (you don't *need* to do this):

Multiply both sides of the equation by 4: $4y - 20 = -7(x + 1)$.
Expand the brackets: $4y - 20 = -7x - 7$.
Rearrange to give the previous result: $4y + 7x = 13$.

An equation of the form $ax + by + c = 0$ can be rearranged into gradient–intercept form provided that $b \neq 0$. This becomes $y = -\frac{a}{b}x - \frac{c}{b}$ and shows that parallel lines may be produced by keeping a and b fixed and allowing c to change.

For example, the lines $2x - 3y = 5$, $2x - 3y = -2$, $4x - 6y = 7$ are all parallel.

> **KEY POINT**
>
> When two straight lines are **perpendicular**, the **product** of their **gradients is** -1.

Example Find the equation of the straight line perpendicular to the line $4x + 3y = 12$ and passing through the point $(2, 5)$.

Rearrange the equation into gradient–intercept form: $y = -\frac{4}{3}x + 4$.
If the gradient of the required line is m then: $m \times -\frac{4}{3} = -1 \Rightarrow m = \frac{3}{4}$.
Using the form $y - y_1 = m(x - x_1)$ gives: $y - 5 = \frac{3}{4}(x - 2)$.

If P has coordinates (x_1, y_1) and Q has coordinates (x_2, y_2) then the mid-point of PQ has coordinates $\left(\dfrac{x_1 + x_2}{2}, \dfrac{y_1 + y_2}{2}\right)$.

For example, the mid-point of the line joining $(-3, 1)$ and $(5, 7)$ is $\left(\dfrac{-3 + 5}{2}, \dfrac{1 + 7}{2}\right) = (1, 4)$.

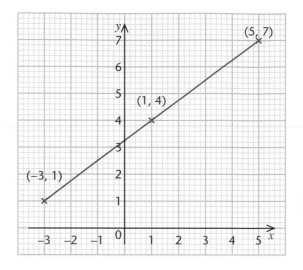

The distance between two points

The distance between the points $A(x_1, y_1)$ and $B(x_2, y_2)$ is given by $\sqrt{(x_1 - x_2)^2 + (y_1 - y_2)^2}$. The result is based on Pythagoras' theorem.

For example, the distance between $(-3, 4)$ and $(2, -8)$ is given by $\sqrt{(-3 - 2)^2 + (4 - (-8))^2} = \sqrt{5^2 + 12^2} = 13$ units.

The equation of a circle

If the centre is at $(0, 0)$, the equation is $x^2 + y^2 = r^2$

The equation of a circle with centre (a, b) and radius r is $(x - a)^2 + (y - b)^2 = r^2$.

This result is based on Pythagoras' theorem.

$(x - a)^2 + (y - b)^2 = r^2$

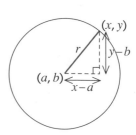

For example, the equation of a circle with centre (4, 0) and radius 5 units is $(x-4)^2 + y^2 = 25$.

> **KEY POINT**
>
> An alternative form of the equation of a circle is $x^2 + y^2 + 2gx + 2fy + c = 0$. The centre of the circle is $(-g, -f)$ and the radius is $\sqrt{g^2 + f^2 - c}$.

For example, the circle with equation $x^2 + y^2 + 6x - 10y + 18 = 0$ has centre $(-3, 5)$ and radius $\sqrt{(-3)^2 + 5^2 - 18} = \sqrt{16} = 4$ units.

You can also show this by completing the square on the x terms and on the y terms as follows:

$$x^2 + 6x + y^2 - 10y + 18 = 0$$
$$(x+3)^2 - 9 + (y-5)^2 - 25 + 18 = 0$$
$$(x+3)^2 + (y-5)^2 = 16$$

> Compare with $(x-a)^2 + (y-b)^2 = r^2$.

The centre is at $(-3, 5)$ and the radius is 4.

Circle properties

You need to remember and be able to use these **circle properties**.

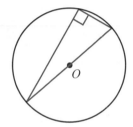

The angle in a **semicircle** is a right angle.

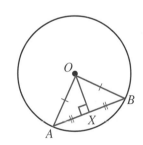

The perpendicular from the centre of a circle to a **chord** bisects the chord.

In the diagram, $AX = XB$.

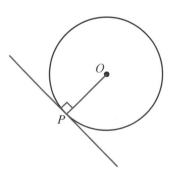

The **tangent** to a circle is perpendicular to the radius at its point of contact.

> The normal to the circle at P is along OP, as it is perpendicular to the tangent.

If you know the gradient of OP, you can work out the gradient of the tangent using the fact that for perpendicular lines, the product of the gradients is -1.

Progress check

The parallel line can be written in the form $3x + 2y = C$. To find C just substitute the coordinates (5, 4) for x and y.

1 The equation of a straight line is $y = 2x - 7$.

   (a) Write down the equation of a parallel line crossing the y-axis at 5.

   (b) Find the equation of a parallel line passing through the point (4, 9).

2 Find the equation of the straight line joining the points (−2, −9) and (1, 3).

3 The equation of a straight line is $3x + 2y = 11$.

   (a) Find the equation of a parallel line passing through the point (5, 4).

   (b) Find the equation of a perpendicular line passing through the point (−2, 6).

4 Find the coordinates of the mid-point of (−9, 3) and (−5, 1).

5 P is the point (−1, −4) and Q is the point (5, −1). Find the equation of the line perpendicular to PQ and passing through its mid-point.

6 A is the point (4, −7) and B is the point (−2, 1).
   Find the distance AB.

7 (a) Find the equation of a circle with centre (−3, 8) and radius 5 units.

   (b) A circle has equation $x^2 + y^2 - 12x + 8y + 43 = 0$.
      Find its centre and radius.

1 (a) $y = 2x + 5$ (b) $y = 2x + 1$
2 $y = 4x - 1$
3 (a) $3x + 2y = 23$ (b) $y - 6 = \frac{2}{3}(x + 2)$ or $3y - 2x - 22 = 0$.
4 (−7, 2)
5 $y = -2x + 1.5$
6 AB = 10 units
7 (a) $(x + 3)^2 + (y - 8)^2 = 25$ (b) (6, −4), 3 units

1.3 Differentiation

After studying this section you should be able to:

LEARNING SUMMARY

- use differentiation to find the exact value of the gradient of a curve
- find the equation of the tangent and normal to a curve at a given point
- determine whether a function is increasing or decreasing in an interval
- locate turning points and use the second derivative to distinguish between them

Gradient function

The **gradient** of a curve changes continuously along its length. Its value at any point P is given by the gradient of the **tangent** to the curve at P. The gradient may be found approximately by drawing.

The exact value of the gradient is found by **differentiation**. This is the **limit** of the gradient of PQ as Q moves towards P.

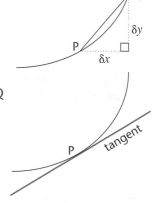

Gradient of curve at P $\approx \dfrac{\delta y}{\delta x}$.

As Q moves towards P, $\delta x \to 0$ and $\dfrac{\delta y}{\delta x} \to \dfrac{\mathrm{d}y}{\mathrm{d}x}$.

> *In Core 1, n will be a positive integer.*

KEY POINT

$\dfrac{\mathrm{d}y}{\mathrm{d}x}$ is the **gradient function** and represents the **derivative** of y with respect to x.

The graph of $y = kx^n$ has gradient function $\dfrac{\mathrm{d}y}{\mathrm{d}x} = nkx^{n-1}$.

Example Find the gradient of the curve $y = 3x^2$ at the point P (5, 75).

$$\frac{\mathrm{d}y}{\mathrm{d}x} = 2 \times 3x^1 = 6x. \quad \text{At P, } x = 5 \text{ so the gradient is } 6 \times 5 = 30.$$

Function notation may also be used for derivatives. If $y = f(x)$ then $\dfrac{\mathrm{d}y}{\mathrm{d}x} = f'(x)$.

The notation is useful for stating some of the rules of differentiation.

KEY POINT

If $y = f(x) \pm g(x)$ then $\dfrac{\mathrm{d}y}{\mathrm{d}x} = f'(x) \pm g'(x)$

Example $f(x) = x^3 + 5x + 2$. Find (a) $f'(x)$ (b) $f'(4)$.

> *Differentiate each term separately. Remember that $x^0 = 1$. When you differentiate a constant, the result is zero.*

(a) $f'(x) = 3x^2 + 5$ (b) $f'(4) = 3 \times 4^2 + 5 = 53$.

You often need to express a function in the right form before you can differentiate it.

Example Differentiate $f(x) = (x - 5)(x + 3)$

You cannot differentiate a product term by term.

The brackets must first be removed and *then* you can differentiate term by term.

$f(x) = (x - 5)(x + 3) = x^2 - 2x - 15$.

$f'(x) = 2x - 2$.

Tangents and normals

You can find the gradient of the tangent to a curve at a point by differentiation. Then you can use the techniques described in the Coordinate Geometry section to find the **equation of the tangent** and the **equation of the normal** at the given point.

Example Find the equation of the tangent and the normal to the curve $y = x^3 - 4x$ at the point (2, 0).

Differentiate the equation of the curve to give $\dfrac{dy}{dx} = 3x^2 - 4$.

The gradient of the tangent at (2, 0) is $3 \times 2^2 - 4 = 8$.
Using $y - y_1 = m(x - x_1)$ gives the equation of the tangent as $y = 8(x - 2)$.

The normal at a point is perpendicular to the tangent.

The gradient of the normal is $-\frac{1}{8}$. Using $y - y_1 = m(x - x_1)$ again gives the equation of the normal as $y = -\frac{1}{8}(x - 2)$. You could rearrange this to give $x + 8y = 2$.

Curve sketching

The gradient function gives information about the behaviour of a curve.

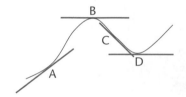

$f'(x) > 0$ at A and the function is **increasing**.

$f'(x) < 0$ at C and the function is **decreasing**.

$f'(x) = 0$ at B and D and the function is neither increasing nor decreasing. These points are called **stationary points**. B is at a **local maximum** and D is at a **local minimum**.

One way to distinguish between the three types of stationary point is to look at the sign of the derivative on either side of the point.

These two types of stationary point are also known as **turning points**.

- At a local maximum the sign of the derivative changes from positive to negative.
- At a local minimum the sign of the derivative changes from negative to positive.

The second derivative

Starting from $\dfrac{dy}{dx} = f'(x)$ and differentiating again gives the **second derivative**. This is written as $\dfrac{d^2 y}{dx^2} = f''(x)$. The second derivative can give information about the nature of any stationary points.

At a stationary point:

- $f''(x) > 0 \implies$ the point is a local minimum
- $f''(x) < 0 \implies$ the point is a local maximum.

But, if $f''(x) = 0$ then this gives no further information.

Example   Find the stationary points of the curve $y = \dfrac{x^3}{3} - \dfrac{3}{2}x^2 + 2x - 1$ and use the second derivative to distinguish between them.

First differentiate the equation, then solve $\dfrac{dy}{dx} = 0$ to locate the stationary points.

$$\frac{dy}{dx} = x^2 - 3x + 2.$$

At a stationary point $\dfrac{dy}{dx} = 0 \implies x^2 - 3x + 2 = 0$

$$\implies (x - 1)(x - 2) = 0$$
$$\implies x = 1 \text{ or } x = 2.$$

When $x = 1$, $y = \frac{1}{3} - \frac{3}{2} + 2 - 1 = -\frac{1}{6}$.

When $x = 2$, $y = \frac{8}{3} - 6 + 4 - 1 = -\frac{1}{3}$.

The stationary points are $(1, -\frac{1}{6})$ and $(2, -\frac{1}{3})$.

In this case, the function is a cubic and so the shape of the curve is known. We should expect the first stationary point to be a local maximum and the second to be a local minimum.

Using the second derivative: $f''(x) = 2x - 3$

so $f''(1) = -1 < 0$ giving a local maximum,

and $f''(2) = 1 > 0$ giving a local minimum as expected.

Progress check

1 Differentiate with respect to x:
   (a) $y = 4x^3 - 7x^2 + 3x - 9$ (b) $y = 6$

2 Differentiate with respect to x:

   $f(x) = (3x + 1)(x - 4)$

3 Find the gradient of the curve $y = (4x - 1)^2$ at the point $(1, 9)$.

Solve the inequality $f'(x) < 0$.

4 Find the values of x for which the function $f(x) = x^3 - 6x^2 + 9x - 7$ is decreasing.

5 Find the equation of: (a) the tangent   (b) the normal
   to the curve $y = x^3 - 2x$ at the point $(2, 4)$.

5 (a) Tangent: $y = 10x - 16$; (b) normal: $x + 10y - 42 = 0$.

4 $1 < x < 3$

3 24

2 $6x - 11$

1 (a) $12x^2 - 14x + 3$ (b) 0

1.4 Integration

After studying this section you should be able to:

- find an indefinite integral of positive integer powers of x
- evaluate a definite integral
- find the area under a curve, and between a line and a curve

Indefinite integration

The idea of a **reverse process** is an important one in many areas of mathematics. In **calculus**, the reverse process of differentiation is **integration** and this turns out to be an extremely important process, in its own right, with many powerful applications.

When you differentiate x^n with respect to x you may think of the process involving two stages:

$$x^n \rightarrow \boxed{\text{multiply by the power}} \rightarrow \boxed{\text{reduce the power by 1}} \rightarrow nx^{n-1}$$

This *suggests* that the reverse process is given by:

In Core 1, n will be a positive integer.

$$\frac{x^{n+1}}{n+1} \leftarrow \boxed{\text{divide by the power}} \leftarrow \boxed{\text{increase the power by 1}} \leftarrow x^n$$

However, this doesn't give the *complete* picture. The reason is that you can differentiate $x^n +$ (any constant) and still obtain nx^{n-1}. You need to take this into account when you reverse the process.

$\int$ is the symbol for integration and dx is used to show that the integration is with respect to the variable x.

> The result is written as $\int x^n \, \mathrm{d}x = \dfrac{x^{n+1}}{n+1} + c$ where c is called the **constant of integration**. Note that $n \neq -1$.
>
> **KEY POINT**

Some examples are:

$$\int x^2 \, \mathrm{d}x = \frac{x^3}{3} + c \qquad \int x^3 \, \mathrm{d}x = \frac{x^4}{4} + c \qquad \int 3 \, \mathrm{d}x = 3x + c$$

Sums and differences of functions are treated in the same way as in differentiation, by dealing with each term separately. The general rules are given below.

Function notation is useful here but you don't need to state the rules every time you use them.

> $\int (f(x) \pm g(x)) \, \mathrm{d}x = \int f(x) \, \mathrm{d}x \pm \int g(x) \, \mathrm{d}x$
>
> $\int af(x) \, \mathrm{d}x = a \int f(x) \, \mathrm{d}x$ (where a is a constant)
>
> **KEY POINT**

For example,

$$\int (3x^2 - 5x + 2) \, \mathrm{d}x = x^3 - \tfrac{5}{2}x^2 + 2x + c.$$

In some situations you have enough information to find the value of the constant.

Example Find the equation of the curve with gradient function $3x^2$ passing through (2, 5).

Using the point (2, 5) gives
$5 = 2^3 + c$
so $c = -3$.

$$\frac{\mathrm{d}y}{\mathrm{d}x} = 3x^2 \Rightarrow y = \int 3x^2 \, \mathrm{d}x \Rightarrow y = x^3 + c. \quad \text{When } x = 2, \ y = 5 \Rightarrow c = -3.$$

The equation of the curve is $y = x^3 - 3$.

Definite integration

An integral of the form $A = \int_a^b f(x)\,\mathrm{d}x$ is a **definite integral**. It has a numerical value and is evaluated as follows:

Substitute the upper limit, then subtract the value when you substitute the lower limit.

A constant of integration is **not** used in a definite integral.

> **KEY POINT**
>
> $$\int_a^b f(x)\,\mathrm{d}x = [g(x)]_a^b = g(b) - g(a)$$

For example $\int_1^4 2x\,\mathrm{d}x = [x^2]_1^4 = 4^2 - 1^2 = 15.$

Area under a curve

For areas *below* the *x*-axis, the definite integral gives a *negative* value.

> **KEY POINT**
>
> The area enclosed by the curve $y = f(x)$, the *x*-axis and the lines $x = a$ and $x = b$ is given by $\int_a^b f(x)\,\mathrm{d}x.$
>
>

Example Find the area under the curve $y = x^2 + 1$ between $x = 1$ and $x = 3$.

$$\text{Area} = \int_1^3 (x^2 + 1)\,\mathrm{d}x = \left[\frac{x^3}{3} + x\right]_1^3$$

$$= \left(\frac{3^3}{3} + 3\right) - \left(\frac{1^3}{3} + 1\right)$$

$$= 9 + 3 - \frac{1}{3} - 1$$

$$= 10\tfrac{2}{3}.$$

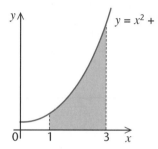

To find the **area between a line and a curve**, find the area under the line and the area under the curve separately and then subtract to find the required area.

Progress check

1 Integrate each of these with respect to x.

(a) x^4 (b) $6x^2 - 4x + 3$ (c) $\dfrac{1}{2}x^2$ (d) $x(x + 5)$ (e) $(x + 3)(x - 8)$.

2 A curve has gradient function $4x^3 + 1$ and passes through the point (1, 9). Find the equation of the curve.

3 Evaluate $\int_0^4 2x^3\,\mathrm{d}x$

4 Calculate the shaded area.

4 $\dfrac{1}{6}$

3 128

2 $y = x^4 + x + 7$

1 (a) $\dfrac{x^5}{5} + c$ (b) $2x^3 - 2x^2 + 3x + c$ (c) $\dfrac{1}{6}x^3 + c$ (d) $\dfrac{x^3}{3} + \dfrac{5x^2}{2} + c$ (e) $\dfrac{x^3}{3} - \dfrac{5x^2}{2} - 24x + c$

Sample questions and model answers

1

(a) Write $2x^2 - 12x + 11$ in the form $a(x + b)^2 + c$.

(b) State the minimum value of $2x^2 - 12x + 11$ and give the value of x where this occurs.

(c) Solve the equation $2x^2 - 12x + 11 = 0$ and express your answer in surd form.

(d) Sketch the graph of $y = 2x^2 - 12x + 11$.

> Start by taking out the factor of 2 common to the first two terms.

> Keep the 11 separate rather than have a fraction inside the brackets.

> Complete the square and simplify.

> The minimum value occurs when the expression in the brackets equals zero.

(a) $2x^2 - 12x + 11 = 2(x^2 - 6x) + 11$

$\qquad\qquad\qquad = 2((x - 3)^2 - 9) + 11$

$\qquad\qquad\qquad = 2(x - 3)^2 - 7.$

(b) The minimum value is -7 and this occurs when $x = 3$.

(c) $2x^2 - 12x + 11 = 0$

$\Rightarrow 2(x - 3)^2 - 7 = 0$

$\Rightarrow (x - 3)^2 = \frac{7}{2} = \frac{14}{4}$

$\Rightarrow x - 3 = \pm\frac{1}{2}\sqrt{14}$

$\Rightarrow x = 3 \pm \frac{1}{2}\sqrt{14}.$

(d)

> When $x = 0$, $y = 11$.

> Use the information you have found to position the curve and label the key points.

Sample questions and model answers *(continued)*

2

A curve has equation $y = (2 - x)(2 + x)$.

The point P(−1, 3) lies on the curve.

(a) Find the gradient of the tangent to the curve at P.

(b) Find the equation of the normal to the curve at P in the form $px + qy = r$.

(c) The normal to the curve at P crosses the y-axis at A and the x-axis at B. Find the area of triangle AOB.

(a) $\quad y = (2 - x)(2 + x)$

$\qquad = 4 - x^2$

$\quad \dfrac{dy}{dx} = -2x.$

Expand the brackets before differentiating.

When $x = -1$, $\dfrac{dy}{dx} = -2(-1) = 2$

The gradient of the tangent at P is 2.

(b) The gradient of the normal at P is $-\frac{1}{2}$.

The product of the gradients must be −1, as they are at right angles to each other.

Equation of the normal at P:

$\quad y - 3 = -\frac{1}{2}(x - (-1))$

$\quad y - 3 = -\frac{1}{2}(x + 1)$

$\quad 2(y - 3) = -(x + 1)$

$\quad 2y - 6 = -x - 1$

$\quad x + 2y = 5$

This is using the equation of a straight line in the form $y - y_1 = m(x - x_1)$.

Write the equation in the required form.

(c) When $x = 0$, $y = \frac{5}{2} \Rightarrow$ A is the point $(0, \frac{5}{2})$

When $y = 0$, $x = 5 \Rightarrow$ B is the point $(5, 0)$

Find where the line intersects the axes.

Area of triangle AOB $= \frac{1}{2} \times 5 \times \frac{5}{2} = \frac{25}{4} = 6\frac{1}{4}$ units^2.

3

A curve has gradient function $2x - 5$. The point P(2, −1) lies on the curve.

Find the equation of the curve.

$\quad \dfrac{dy}{dx} = 2x - 5$

$\quad y = \displaystyle\int (2x - 5) \, dx$

$\qquad = x^2 - 5x + c$

Remember the integration constant c here.

When $x = 2$, $y = -1$

$\Rightarrow -1 = 2^2 - 5(2) + c$

$\qquad c = -1 - 4 + 10 = 5$

Use the fact that P lies on the curve to find the value of c.

The curve has equation $y = x^2 - 5x + 5$.

Practice examination questions

1 (a) Write $x^2 - 4x + 1$ in the form $(x + a)^2 + b$.

(b) Solve $x^2 - 4x + 1 = 0$ and express your answer in surd form.

(c) State the coordinates of the lowest point on the graph of $y = x^2 - 4x + 1$.

(d) Sketch the graph.

2 A function $f(x)$ is defined by:

$$f(x) = x^3 + 4x^2 - 3x - 18$$

for all real values of x.

(a) Given that $(x - 2)$ is a factor of $f(x)$, factorise $f(x)$ completely.

(b) Sketch the graph of $y = f(x)$.

(c) Sketch the graphs of $y = f(x) + 10$ and $y = f(x - 2)$.

3 Find the coordinates of the points A and B at the intersection of these graphs.

4 Solve the inequality

$$3x^2 - 8x - 7 < 2x^2 - 3x - 11.$$

5 (a) The diagram shows a straight line passing through the points A(−1, −5) and B(4, 1). Find its equation in the form $ax + by + c = 0$.

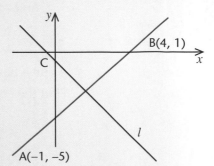

(b) A second line, l, passes through the mid-point of A and B at right angles to AB.

Find its equation in the form $ax + by + c = 0$.

(c) The line l crosses the y-axis at C. Find the coordinates of C.

Practice examination questions *(continued)*

6 The diagram shows two lines l and m at right angles to each other. The equation of the line m is $3x + 2y = 6$. The line l passes through the point with coordinates (5, 4).
l and m cross the x-axis at A and B respectively.

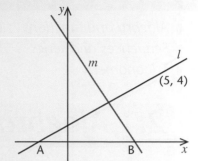

   (a) Find the equation of the line l.

   (b) Show that AB = 3.

7 (a) Find the coordinates of the stationary points on the curve
$$y = 2x^3 - 15x^2 - 36x + 10.$$

   (b) Use the second derivative to determine the nature of the stationary points found in (a).

   (c) Find the set of values of x for which $2x^3 - 15x^2 - 36x + 10$ is a decreasing function of x.

8 The diagram shows a straight line crossing the curve $y = (x - 1)^2 + 4$ at the points P and Q. It passes through (0, 7) and (7, 0).

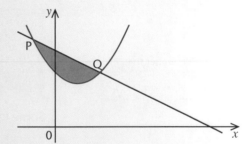

   (a) Write down the equation of the straight line through P and Q.

   (b) Find the coordinates of P and Q.

   (c) Calculate the value of the shaded area.

9 (a) Given that $(x + 2)$ is a factor of $f(x) = x^3 - 4x^2 - 3x + k$, use the factor theorem to find the value of k.

   (b) Factorise $f(x)$ completely and sketch the graph of $y = f(x)$.

   (c) Solve the inequality $f(x) \geqslant 0$.

   (d) Find the remainder when $f(x)$ is divided by $(x + 1)$.

Chapter 2
Core 2 Pure Mathematics

The following topics are covered in this chapter:

- Algebra and functions
- Sequences and series
- Trigonometry
- Differentiation
- Integration

2.1 Algebra and functions

After studying this section you should be able to:

- work with indices
- understand exponential functions and sketch their graphs
- understand the log laws and use them to solve equations of the form $a^x = b$
- use transformations to sketch graphs of related functions

LEARNING SUMMARY

Indices

You need to know these basic rules and be able to apply them.

$$a^m \times a^n = a^{m+n} \qquad a^m \div a^n = a^{m-n} \qquad (a^m)^n = a^{mn} \qquad a^{\frac{1}{n}} = \sqrt[n]{a}$$

$$a^{\frac{m}{n}} = (a^m)^{\frac{1}{n}} = (a^{\frac{1}{n}})^m \qquad a^{-n} = \frac{1}{a^n} \qquad (ab)^n = a^n b^n \qquad \left(\frac{a}{b}\right)^n = \frac{a^n}{b^n}$$

> Remember that, in general $(a+b)^n \neq a^n + b^n$ and $(a-b)^n \neq a^n - b^n$.

Some important special cases are: $\quad a^1 = a \qquad a^{\frac{1}{2}} = \sqrt{a}$

and $\quad a^0 = 1 \qquad a^{-1} = \dfrac{1}{a} \quad$ provided $a \neq 0$.

For example $\quad 9^{\frac{1}{2}} = \sqrt{9} = 3 \quad$ and $\quad 16^{\frac{3}{2}} = (16^{\frac{1}{2}})^3 = 4^3 = 64$.

More graphs

The graph of some new function can often be obtained from the graph of a known function by applying a transformation. A summary of the standard transformations is given in the table.

Known function	New function	Transformation
$y = f(x)$	$y = f(x) + a$	Translation through a units parallel to y-axis.
	$y = f(x - a)$	Translation through a units parallel to x-axis.
	$y = af(x)$	One-way stretch with scale factor a parallel to the y-axis.
	$y = f(ax)$	One-way stretch with scale factor $\frac{1}{a}$ parallel to the x-axis.

> You may need to apply a combination of transformations in some cases.

Example   The diagram shows the graph of a function, $y = f(x)$ for $1 \leqslant x \leqslant 3$.

Use the same axes to show:
(a) $y = f(x) + 1$
(b) $y = f(x + 1)$
(c) $y = 2f(x)$
(d) $y = f(2x)$

(a)

(b)

(c)

$y = 2f(x)$

(d)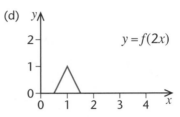

$y = f(2x)$

Exponential functions

An **exponential function** is one where the variable is a power or exponent. For example, any function of the form $f(x) = a^x$, where a is a constant, is an exponential function.

The diagram shows the graphs of $y = 2^x$ and $y = 10^x$.

> Exponential functions are often used to represent, or model, patterns of:
> • growth when $a > 1$
> • decay when $a < 1$.

All graphs of the form $y = a^x$ cross the y-axis at (0, 1). Each graph has a different gradient at this point and the value of the gradient depends on a.

In Core 3 a special exponential function is studied, with a gradient 1 at (0, 1).

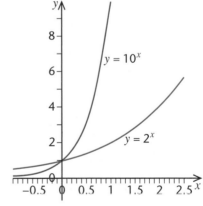

Logarithms

Exponential functions are related to **logarithmic functions** as follows:

$$y = \log_a x \Leftrightarrow x = a^y$$

a is called the base of the logarithm.

> $a > 0$ and $a \neq 1$.

> Your calculator has a key for logs to the base 10, probably labelled log.

The **logarithm** (log) to a given **base** of a number is the **power** to which the base must be raised to equal the number.

For example, $10^2 = 100$, so $\log_{10} 100 = 2$.

It is useful to remember the following are true for any base a:

$\log_a a = 1$, since $a^1 = a$

$\log_a 1 = 0$, since $a^0 = 1$.

The graph of $y = \log_a x$ is a reflection of $y = a^x$ in the line $y = x$.

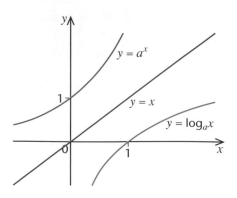

The log laws

There are three laws of logarithms that you need to know. The same laws apply in any base and so no particular base is stated:

- $\log a + \log b = \log ab$

- $\log a - \log b = \log\left(\dfrac{a}{b}\right)$

- $n \log a = \log a^n$.

You can use these laws to simplify expressions involving logs and to solve **exponential equations**, i.e. equations where the unknown value is a power.

Example Express $\log x + 3 \log y$ as a single logarithm.

$$\log x + 3 \log y = \log x + \log y^3 \qquad \text{(using the third law)}$$
$$= \log xy^3 \qquad \text{(using the first law)}$$

Example Solve the equation $5^x = 30$.

Taking logarithms of both sides gives:

$$\log 5^x = \log 30$$
$$\Rightarrow x \log 5 = \log 30$$
$$\Rightarrow x = \frac{\log 30}{\log 5} = 2.113 \text{ to 4 s.f.}$$

This works in any base.

Progress check

1 Simplify each of these without using a calculator:

(a) $25^{\frac{3}{2}}$ (b) $9^{-\frac{1}{2}}$.

2 The diagram shows the graph of $y = f(x)$ for $0 \leqslant x \leqslant 2$.
Sketch the graph of:

(a) $y = f(x - 2)$
(b) $y = f(2x)$.

3 Solve the equation $4^x = 100$ and give your answer to 4 s.f.

4 Write each of the following as a single logarithm.
(a) $\log_a x + 3\log_a y - \frac{1}{2}\log_a z$
(b) $\log_{10} x - 1$

5 Evaluate the following:
(a) $\log_2 8$
(b) $\log_9 3$
(c) $2\log_4 2$

6 Solve $3^{2x} = 50$

6 1.78 (3 s.f.)

5 (a) 3 (b) $\frac{1}{2}$ (c) 1

4 (a) $\log_a\left(\dfrac{xy^3}{\sqrt{z}}\right)$ (b) $\log_{10}\left(\dfrac{x}{10}\right)$

3 $x = 3.322$

2

1 (a) 125 (b) $\frac{1}{3}$

2.2 Sequences and series

After studying this section you should be able to:

- understand the different types of sequence, including the notation and formulae used to describe them
- find the limit of a convergent sequence
- recognise arithmetic and geometric progressions and calculate the sum of their series
- determine whether a geometric series with an infinite number of terms has a finite sum and calculate its value when it exists
- use the binomial expansion for positive integer powers of n

Sequences and series

A list of numbers in a particular order, that follow some rule for finding later values, is called a **sequence**. Each number in a sequence is called a **term**, and terms are often denoted by $u_1, u_2, u_3, ..., u_n, ...$.

One way to define a sequence is to give a formula for the nth term such as $u_n = n^2$. Substituting the values $n = 1, 2, 3, 4, ...$ produces the sequence 1, 4, 9, 16, ... and the value of any particular term can be calculated by substituting its position number into the formula. For example, in this case, the 50th term is $u_{50} = 50^2 = 2500$.

Another way to define a sequence is to give a starting value together with a rule that shows the connection between successive terms. This is sometimes called a **recursive definition**. For example $u_1 = 5$ and $u_{n+1} = 2u_n$ defines the sequence 5, 10, 20, 40, ... The rule $u_{n+1} = 2u_n$ is an example of a **recurrence relation**.

Two special sequences are the **arithmetic progression** (A.P.) and the **geometric progression** (G.P.).

In an A.P. successive terms have a **common difference**, e.g. 1, 4, 7, 10, ... The first term is denoted by a and the common difference is d. With this notation, the definition of an A.P. may now be given as $u_1 = a$, $u_{n+1} = u_n + d$.

The terms of an A.P. take the form a, $a + d$, $a + 2d$, $a + 3d$, ... and the nth term is given by $u_n = a + (n - 1)d$.

In a G.P. successive terms are connected by a **common ratio**, e.g. 3, 6, 12, 24, ... The definition of a G.P. may be given as $u_1 = a$, $u_{n+1} = ru_n$. where r is the common ratio.

The terms of a G.P. take the form $a, ar, ar^2, ar^3, ...$ and the nth term is given by $u_n = ar^{n-1}$.

A **series** is formed by adding together the terms of a sequence. The use of sigma notation can greatly simplify the way that series are written. For example, the series $1^2 + 2^2 + 3^2 + ... + n^2$ may be written as $\sum_{i=1}^{n} i^2$. The sum of the first n terms of a series is often denoted by S_n and so $S_n = u_1 + u_2 + u_3 + ... + u_n = \sum_{i=1}^{n} u_i$.

Infinite sequences

In an arithmetic sequence such as 3, 5, 7, 9, ... the terms keep on getting bigger and, by taking the sequence far enough, they can be made bigger than any given value. Such a sequence is said to be **divergent**.

In a sequence such as $u_{n+1} = \dfrac{u_n}{2} + 5$, the terms approach a particular value ever more closely as $n \to \infty$. This value is called the **limit** of the sequence and the sequence is said to be **convergent**.

If we use L to represent the value of the limit then, in this case, $L = \dfrac{L}{2} + 5$.

This equation may be solved to give $L = 10$.

This shows that the sequence has a limit of 10 regardless of the starting value.

In general, if a convergent sequence is given by $x_{n+1} = f(x_n)$, then the limit, L, is found by solving $L = f(L)$.

Arithmetic series

The **sum of an arithmetic series** is given by
$$S_n = a + (a+d) + (a+2d) + \ldots + (a + (n-1)d).$$
This may also be written as $S_n = l + (l-d) + (l-2d) + \ldots + (l - (n-1)d),$
where l is the last term.

Adding the two versions gives $2S_n = (a+l) + (a+l) + (a+l) + \ldots (a+l)$
$$= n(a+l).$$
$$S_n = \dfrac{n}{2}(a+l).$$

You need to know how to establish the general result.

> **KEY POINT**
>
> The sum of the first n terms of an arithmetic series is: $S_n = \dfrac{n}{2}(a+l)$.
>
> Substituting $l = a + (n-1)d$ gives the alternative form of the result:
> $$S_n = \dfrac{n}{2}(2a + (n-1)d).$$

A special case is the sum of the first n natural numbers:

$$1 + 2 + 3 + \ldots + n = \dfrac{n}{2}(n+1).$$

For example, the sum of the first 100 natural numbers is $\frac{100}{2}(100+1) = 5050$.

When finding the sum of an A.P. you need to select the most appropriate version of the formula to suit the information.

Example   Find the sum of the first 50 terms of the series $15 + 18 + 21 + 24 + \ldots$.

In this series, $a = 15$, $d = 3$ and $n = 50$.

Using $S_n = \dfrac{n}{2}(2a + (n-1)d)$ gives $S_{50} = \dfrac{50}{2}(2 \times 15 + 49 \times 3) = 4425$.

Geometric series

The **sum of a G.P.** is given by $S_n = a + ar + ar^2 + ar^3 + \ldots + ar^{n-1}$.

Multiplying throughout by r gives $rS_n = ar + ar^2 + ar^3 + \ldots + ar^{n-1} + ar^n$.

Subtracting gives: $S_n - rS_n = a - ar^n$
$$\Rightarrow S_n(1-r) = a(1 - r^n)$$

You need to know how to establish this result.

> **KEY POINT**
>
> So the sum of the first n terms of a geometric series is $S_n = \dfrac{a(1-r^n)}{1-r}$.

If $r > 1$ then it is more convenient to use the result in the form $S_n = \dfrac{a(r^n - 1)}{r - 1}$.

Example Find the sum of the first 20 terms of the series $8 + 12 + 18 + 27 + \ldots$ to the nearest whole number.

In this series, $a = 8$, $r = 1.5$ and $n = 20$.

This gives $S_{20} = \dfrac{8(1.5^{20} - 1)}{1.5 - 1} = 53188.107\ldots = 53188$ to the nearest whole number.

> Provided that $|r| < 1$, the sum of a geometric series converges to $\dfrac{a}{1 - r}$ as n tends to infinity. This is known as the **sum to infinity** of a geometric series.
>
> KEY POINT

For example, $1 + \dfrac{1}{2} + \dfrac{1}{4} + \dfrac{1}{8} + \ldots = \dfrac{1}{1 - \frac{1}{2}} = 2$.

Binomial expansion

An expression which has two terms, such as $a + b$ is called a **binomial**. The expansion of something of the form $(a + b)^n$ is called a **binomial expansion**. When n is a positive integer:

The situation where n is not a positive integer can be dealt with but you don't need it for this component.

$$(a + b)^n = a^n + na^{n-1}b + \frac{n(n-1)}{2!} a^{n-2}b^2 + \frac{n(n-1)(n-2)}{3!} a^{n-3}b^3 + \ldots + b^n.$$

This expansion may appear complicated but it starts to look simpler if you follow the patterns from one term to the next:

Remember that $a^0 = 1$ when $a \neq 0$.

- Starting with a^n, the power of a is reduced by 1 each time until the last term, b^n, which is the same as $a^0 b^n$.

- The power of b is increased by 1 each time. Notice that the first term, a^n, is the same as $a^n b^0$ and that the powers of a and b always add up to n in each term.

- It's worth remembering the first three coefficients: 1, n and $\dfrac{n(n-1)}{2!}$ then the rest follow the same pattern, giving $\dfrac{n(n-1)(n-2)}{3!}$, $\dfrac{n(n-1)(n-2)(n-3)}{4!}$ and so on.

Pascal's triangle: each row starts and ends with 1. Every other value is found by adding the pair of numbers immediately above it in the pattern.

```
        1    4    6
   1    5    10
```

The coefficients in the expansion of $(a + b)^n$ also follow the pattern given by the row of **Pascal's triangle** that starts with 1 n …

So the coefficients in the expansion of $(a + b)^4$ for example are 1, 4, 6, 4 and 1.

```
                    1
                1       1
            1       2       1
        1       3       3       1
    1       4       6       4       1
1       5       10      10      5       1
```

It follows that
$(a + b)^4 = a^4 + 4a^3b + 6a^2b^2 + 4ab^3 + b^4.$

Variations on this result can be obtained by substituting different values for a and b. Some examples are:

In this case, $a = 1$ and $b = x$.

$$(1 + x)^4 = 1 + 4x + 6x^2 + 4x^3 + x^4.$$

and

For this one, $a = 1$ and $b = -2x$.

$$(1 - 2x)^4 = 1 + 4(-2x) + 6(-2x)^2 + 4(-2x)^3 + (-2x)^4$$
$$= 1 - 8x + 24x^2 - 32x^3 + 16x^4.$$

The notation $\binom{n}{r}$ is often used to stand for the expression $\dfrac{n(n-1)\ldots(n-r+1)}{r!}$.

So, for example $\binom{n}{3} = \dfrac{n(n-1)(n-2)}{3!}$.

Using this notation, the binomial expansion may be written as:

$$(a+b)^n = a^n + \binom{n}{1}a^{n-1}b + \binom{n}{2}a^{n-2}b^2 + \binom{n}{3}a^{n-3}b^3 + \ldots + b^n.$$

- It's useful to recognise that the term involving b^r takes the form $\binom{n}{r}a^{n-r}b^r$.

- For positive integer values $\binom{n}{r}$ has the same value as ^nC_r and you may find that your calculator will work this out for you.

- When n is small and the full binomial expansion is required, the simplest way to find the coefficients is to use Pascal's triangle. For larger values of n it may be simpler to use the formula, particularly if only some of the terms are required.

Example   Expand $(1+3x)^{10}$ in ascending powers of x up to and including the fourth term.

$$(1+3x)^{10} = 1 + \binom{10}{1}(3x) + \binom{10}{2}(3x)^2 + \binom{10}{3}(3x)^3 + \ldots$$

$$= 1 + 10(3x) + 45(9x^2) + 120(27x^3) + \ldots$$

$$= 1 + 30x + 405x^2 + 3240x^3 + \ldots.$$

Example   Find the coefficient of the x^7 term in the expansion of $(3-2x)^{15}$

Notice that the coefficient of x^7 is not totally determined by the value of $\binom{15}{7}$.

The term involving x^7 is given by $\binom{15}{7}(3)^8(-2x)^7$

$$= 6435 \times 6561 \times -128x^7,$$

and so the required coefficient is $-5\,404\,164\,480$.

Progress check

1   Write down the first five terms of the sequence given by:
   (a) $u_n = 2n^2$ (b) $u_1 = 10$, $u_{n+1} = 3u_n + 2$.

2   Find the 20th term of an A.P. with first term 7 and common difference 5.

3   Find the sum of the first 1000 natural numbers.

4   Find the sum of the first 20 terms of $12 + 15 + 18.75 + \ldots$ to 4 s.f.

5   Explain why the series $10 + 9 + 8.1 + 7.29 + \ldots$ is convergent and find the value of its sum to infinity to 5 s.f.

3 500 500   4 4115   5 $|r| = 0.9 < 1$, 100.

2 102

1 (a) 2, 8, 18, 32, 50 (b) 10, 32, 98, 296, 890

2.3 Trigonometry

After studying this section you should be able to:

- state the exact values of the sine, cosine and tangent of special angles
- understand the properties of the sine, cosine and tangent functions
- sketch the graphs of trigonometric functions
- use the sine and cosine rules and the formula for the area of a triangle
- understand radian measure and use it to find arc lengths and areas of sectors
- solve trigonometric equations in a given interval
- understand and use trigonometric identities

LEARNING SUMMARY

Sine, cosine and tangent of special angles

The sine, cosine and tangent of 30°, 45°, and 60° may be expressed exactly as shown below. The results are based on Pythagoras' theorem.

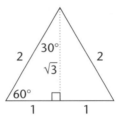

Knowledge of the exact values is not a requirement of the AQA specification but an awareness of them may be useful.

$$\sin 30° = \frac{1}{2} \qquad \sin 60° = \frac{\sqrt{3}}{2} \qquad \sin 45° = \frac{1}{\sqrt{2}}$$

$$\cos 30° = \frac{\sqrt{3}}{2} \qquad \cos 60° = \frac{1}{2} \qquad \cos 45° = \frac{1}{\sqrt{2}}$$

$$\tan 30° = \frac{1}{\sqrt{3}} \qquad \tan 60° = \sqrt{3} \qquad \tan 45° = 1.$$

Trigonometric graphs

You need to be able to sketch the graphs of the sine, cosine and tangent functions and to know their special properties.

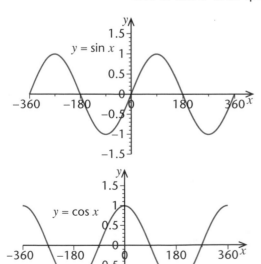

$\sin x$ is defined for any angle and always has a value between −1 and 1. It is a **periodic function** with period 360°.

The graph has **rotational symmetry** of order 2 about every point where it crosses the x-axis.

It has **line symmetry** about every vertical line passing through a vertex.

$\cos x \equiv \sin(x + 90°)$ so the graph of $y = \cos x$ can be obtained by translating the sine graph 90° to the left.

It follows that $\cos x$ is also a periodic function with period 360° and has the corresponding symmetry properties.

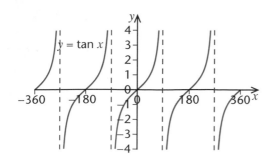

$$\tan x \equiv \frac{\sin x}{\cos x}.$$

$\tan x$ is undefined whenever $\cos x = 0$ and approaches $\pm\infty$ near these values. It is a periodic function with period $180°$.

The graph has rotational symmetry of order 2 about $0°$, $\pm 90°$, $\pm 180°$, $\pm 270°$, ...

The graphs of more complex trigonometric functions can often be produced by applying transformations to one of the basic graphs.

Example
Sketch the graphs of $y = \sin(x)$ and $y = \sin(2x)$ for $0° \leqslant x \leqslant 360°$ on the same axes.

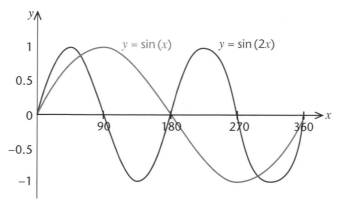

The transformation that changes the graph of $y = \sin(x)$ into the graph of $y = \sin(2x)$ is a one-way stretch in the x direction with scale factor $\frac{1}{2}$. The effect of the transformation is that the graph of $y = \sin(2x)$ goes through its cycle twice as quickly.

Area of a triangle

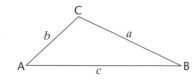

The same formula applies whether the angle is acute or obtuse.

Area of triangle $= \frac{1}{2} ab \sin C$

In both the diagrams, $b \sin C$ represents the height of the triangle.

The sine and cosine rules

A, B and C can be at any vertex. The opposite sides are then labelled as a, b, c.

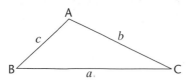

Sine rule: $\dfrac{a}{\sin A} = \dfrac{b}{\sin B} = \dfrac{c}{\sin C}$

Cosine rule: $a^2 = b^2 + c^2 - 2bc \cos A$

This may be rearranged to find an angle

$$\cos A = \frac{b^2 + c^2 - a^2}{2bc}$$

Radians

Angles can also be measured in **radians** and this makes it much easier to deal with trigonometric functions when using **calculus**.

1 radian $\approx 57.3°$. This may be written as $1^c \approx 57.3°$. However, the symbol for radians is not normally written when the angle involves π.

The following results are useful to remember:

$$\pi = 180°, \frac{\pi}{2} = 90°, \frac{\pi}{4} = 45°, \frac{\pi}{3} = 60°, \frac{\pi}{6} = 30°.$$

All the results that you know in degrees also apply in radians, e.g. $\sin \dfrac{\pi}{6} = 0.5$.

Arc length and sector area

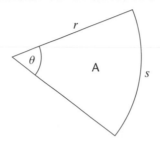

With θ in radians:

- Length of arc of sector is given by
  $s = r\theta$.

- Area of sector is given by
  $A = \frac{1}{2}r^2\theta$.

Solving trigonometric equations

You may be required to solve trigonometric equations in a variety of forms.

Example   Solve $\sin(2x - 30°) = 0.6$ for $0° \leqslant x \leqslant 360°$.

> This is the principal value given by the calculator.

First find a value of $2x - 30°$.

One value of $2x - 30°$ is $\sin^{-1}(0.6) = 36.86...°$.
You need to adjust the interval to find where the values of $2x - 30°$ must lie.

$$0° \leqslant x \leqslant 360° \Rightarrow 0° \leqslant 2x \leqslant 720° \Rightarrow -30° \leqslant 2x - 30° \leqslant 690°.$$

Now look at the graph of the sine function in this interval.

*This is the graph of $y = \sin X$ for $-30° \leqslant X \leqslant 690°$ where $X = 2x - 30°$.
In other words, draw the sine curve as normal for the interval and then interpret the variable as $2x - 30°$.*

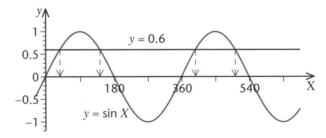

The graph shows that there are four values of $2x - 30°$ in the interval that make the equation work.

Using the symmetry of the curve, the values of $2x - 30°$ are:

$36.86...°, 180° - 36.86...°, 360° + 36.86...°, 540° - 36.86...°$.

This gives:

$2x - 30° = 36.86...° \Rightarrow x = 33.4°$
$2x - 30° = 143.14...° \Rightarrow x = 86.6°$

$2x - 30° = 396.86...° \Rightarrow x = 213.4°$
$2x - 30° = 503.14...° \Rightarrow x = 266.6°$
So $x = 33.4°, 86.6°, 213.4°, 266.6°$ (1 d.p.).

Using trigonometric identities

The following trigonometric relationships are **identities**. This means that they are true for all values of x.

They are very important when simplifying expressions and solving equations and should be learnt.

$$\tan x = \frac{\sin x}{\cos x}$$

KEY POINT

Example
Solve the equation $\sin x = \sqrt{3}\cos x$, for values of x between $0°$ and $360°$.

> This is allowed since $\cos x = 0$ is not a solution.

Divide each side by $\cos x$.

$$\frac{\sin x}{\cos x} = \sqrt{3}$$

$$\Rightarrow \tan x = \sqrt{3}$$

$$x = 60° \text{ or } 240°$$

> The graph of $y = \tan x$ repeats every $180°$, so to find the second solution, add $180°$

$$\sin^2 x + \cos^2 x = 1$$

KEY POINT

This is sometimes known as the Pythagorean trigonometric identity.

Example Solve the equation $\sin^2 x - \cos^2 x - \sin x = 1$ for $-\pi \leqslant x \leqslant \pi$.

> You need to recognise that this is a quadratic in $\sin x$.
> It has to be rearranged in order to use the quadratic formula.

Replacing $\cos^2 x$ with $1 - \sin^2 x$ gives $\sin^2 x - (1 - \sin^2 x) - \sin x = 1$

$$\Rightarrow 2\sin^2 x - \sin x - 2 = 0.$$

> $-1 \leqslant \sin x \leqslant 1$ for all values of x.

Using the formula gives $\sin x = \dfrac{1 \pm \sqrt{1 + 16}}{4} = \dfrac{1 \pm \sqrt{17}}{4}$

$$\Rightarrow \sin x = 1.28 \ldots \text{ (no solutions) or } \sin x = -0.78077 \ldots.$$

> Remember to set your calculator to radians mode.

Using $\sin^{-1}$ gives $\qquad x = -0.8959^c$ to 4 d.p.

Another solution in the interval $-\pi \leqslant x \leqslant \pi$ is $-\pi + 0.8959^c$ giving $x = -2.2457^c$ to 4 d.p.

The diagram shows there are no other solutions in this interval.

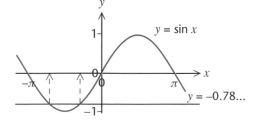

Progress check

1 OAB is a sector of a circle, centre O, radius 4 cm.

Angle AOB = 0.5 radians.

(a) Find the perimeter of sector AOB.

(b) Find the length of the chord AB.

(c) Find the area of the sector OAB.

(d) Find the area of the triangle OAB.

(e) Find the area of the shaded segment.

2 Describe the transformations needed to turn the graph of $y = \cos x$ into the graph of:

(a) $y = \cos 2x$ (b) $y = 3 \cos x$

3 Find the value of x and θ.

4 Find the value of:

(a) $\sin \dfrac{\pi}{4}$ (b) $\cos \dfrac{\pi}{3}$ (c) $\tan \dfrac{\pi}{6}$.

5 Solve $8 \sin x = 3 \cos x$ for $-360° \leqslant x \leqslant 360°$.

6 Solve $\cos(3x + 20°) = 0.6$ for $0° \leqslant x \leqslant 180°$.

7 Solve $2 \sin^2 x + \cos x = 1$ for $-\pi \leqslant x \leqslant \pi$.

2.4 Differentiation

After studying this section you should be able to:

- *recall differentiation methods and applications required in Core 1*
- *extend your knowledge to differentiating negative and rational powers of x*

All the techniques learnt in Core 1 are extended to include all rational n and a reminder is included below.

> If $y = kx^n$, then $\dfrac{dy}{dx} = nkx^{n-1}$ for all rational values of n.

You now have to differentiate expressions when n is not necessarily a positive integer. You may have to change the expression into index form first.

Example 1 Differentiate $y = x^{\frac{3}{2}} + \dfrac{3}{x^2}$

Re-write as $y = x^{\frac{3}{2}} + 3x^{-2}$ and differentiate term by term to give

$$\frac{dy}{dx} = \frac{3}{2}x^{\frac{1}{2}} + (-2) \times 3x^{-3}$$

$$= \frac{3}{2}\sqrt{x} - \frac{6}{x^3}$$

Example 2 Differentiate $y = x\sqrt{x}$

Re-write as $y = x^{\frac{3}{2}}$ giving

$$\frac{dy}{dx} = \frac{3}{2}\sqrt{x}$$

Progress check

1   Find $\dfrac{dy}{dx}$ in each of the following

(a) $y = x^{\frac{5}{3}} - \dfrac{1}{x}$

(b) $y = x^3\sqrt{x}$

2   A curve has equation $\dfrac{1}{x} + 16x$.

(a) Find the co-ordinates of the stationary points on the curve and determine their nature.

(b) Find the equation of the tangent to the curve at the point where $x = 0.5$

2 (a) $(-0.25, -8)$, $(0.25, 8)$ (b) $y = 12x + 4$

1 (a) $\frac{5}{3}x^{\frac{2}{3}} + \frac{1}{x^2}$

(b) $\frac{7}{2}x^{\frac{5}{2}}$ or $\frac{7}{2}\sqrt{x^5}$ or $\frac{7}{2}x^2\sqrt{x}$

2.5 Integration

After studying this section you should be able to:

- recall integration methods and applications required in Core 1
- extend your knowledge to integrating negative and rational powers of x ($n \neq -1$)
- use the trapezium rule to find an approximate value of an integral

Rational powers

Note that functions in the Core 2 module contain values of n that are rational and not necessarily a positive integer, as in the following examples.

Example 1

$$\int (x^{\frac{3}{2}} + 2x^{-\frac{1}{2}})dx = \frac{x^{\frac{5}{2}}}{\frac{5}{2}} + 2\frac{x^{\frac{1}{2}}}{\frac{1}{2}} + c$$

$$= \frac{2}{5}x^{\frac{5}{2}} + 4x^{\frac{1}{2}} + c$$

Example 2

Write the expression in index form first.

$$\int \frac{x+2}{\sqrt{x}}\,dx = \int (x^{\frac{1}{2}} + 2x^{-\frac{1}{2}})dx$$

$$= \frac{2}{3}x^{\frac{3}{2}} + 4x^{\frac{1}{2}} + c$$

Progress check

1 Find $\int \frac{x^2 - 1}{\sqrt{x}}\,dx$

2 Evaluate $\int_1^2 (1 + \frac{3}{x^2})dx$.

3 Find the area enclosed between the curve $y = \frac{1}{\sqrt{x}}$, the x-axis and the lines $x = 1$ and $x = 4$.

3 2

2 2.5

1 $\frac{2}{5}x^{\frac{5}{2}} - 2x^{\frac{1}{2}} + c$

Numerical integration – trapezium rule

The value of $A = \int_a^b f(x)\mathrm{d}x$ represents the area under the graph of $y = f(x)$ between $x = a$ and $x = b$. You can find an approximation to this value using the **trapezium rule**. This is especially useful if the function is difficult to integrate.

The area under the curve between a and b may be divided into n strips of equal width h.

It follows that $h = \dfrac{b-a}{n}$.

Each strip is approximately a trapezium and so the total area is approximately:

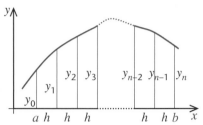

$$\tfrac{1}{2}h(y_0 + y_1) + \tfrac{1}{2}h(y_1 + y_2) + \tfrac{1}{2}h(y_2 + y_3) + \ldots + \tfrac{1}{2}h(y_{n-2} + y_{n-1}) + \tfrac{1}{2}h(y_{n-1} + y_n).$$

To increase the accuracy, use more strips.

KEY POINT

This simplifies to give the formula known as the **trapezium** rule:

$$\int_a^b y\,\mathrm{d}x = \tfrac{1}{2}h\{(y_0 + y_n) + 2(y_1 + y_2 + \ldots + y_{n-1})\} \text{ where } h = \frac{b-a}{n}$$

Example

Use the trapezium rule with six ordinates (five strips) to estimate the value of $\int_1^2 2^x\,\mathrm{d}x$.

It's useful to tabulate the information:

Work to a greater level of accuracy in your table than you intend to give in your final answer.

x	1	1.2	1.4	1.6	1.8	2
2^x	2	2.2974	2.6390	3.0314	3.4822	4
	y_0	y_1	y_2	y_3	y_4	y_5

$$h = \frac{2-1}{5} = 0.2$$

In this case, the working was done to 5 s.f. and the final answer is given to 3 s.f.

So, $\int_1^2 2^x\,\mathrm{d}x \approx 0.1\{(2 + 4) + 2(2.2974 + 2.6390 + 3.0314 + 3.4822)\}$

$$= 2.89 \quad \text{to} \quad 3 \text{ s.f.}$$

Increasing gradient: overestimate

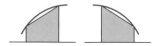

Decreasing gradient: underestimate

If the gradient of the graph is increasing over the interval then the trapezium rule will give an over-estimate of the area. If the gradient of the graph is decreasing over the interval then it will give an under-estimate of the area.

A graphic calculator with a numerical integration function gives the value of the area as 2.8853901

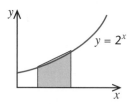

$y = 2^x$

In this case, the gradient is increasing over the interval and so the trapezium rule gives an over-estimate of the area.

Progress check

1 Use the trapezium rule with six ordinates (five strips) to estimate $\int_1^2 \frac{1}{x}\, dx$.

State whether the value obtained is an over-estimate or an under-estimate.

1 0.69563 (5 d.p.) Over-estimate

Sample questions and model answers

1

(a) Write down the exact value of 3^{-3}.

(b) Simplify $\dfrac{(4x^4)^{\frac{3}{2}}}{2\sqrt{x} \times x\sqrt{x}}$.

'Write down' means that you don't need to show any working to get the marks. But you might still find it useful as a way to clarify your thoughts.

Remember that $a^{-n} = \dfrac{1}{a^n}$

(a) $3^{-3} = \dfrac{1}{3^3} = \dfrac{1}{27}$.

$4^{\frac{3}{2}} = (\sqrt{4})^3 = 2^3 = 8$
$(x^4)^{\frac{3}{2}} = x^{4 \times \frac{3}{2}} = x^6$.

(b) $\dfrac{(4x^4)^{\frac{3}{2}}}{2\sqrt{x} \times x\sqrt{x}} = \dfrac{4^{\frac{3}{2}}(x^4)^{\frac{3}{2}}}{2x^2}$

$= \dfrac{8x^6}{2x^2} = 4x^4$.

2

Find the values of x, where $0 \leqslant x \leqslant 2\pi$, that satisfy the equation

$$\tan x = 2 \sin x.$$

Use the identity
$\tan x = \dfrac{\sin x}{\cos x}$

$\tan x = 2 \sin x$

$\dfrac{\sin x}{\cos x} = 2 \sin x$

Multiply both sides by $\cos x$.

$\sin x = 2 \sin x \cos x$

$\sin x - 2 \sin x \cos x = 0$

Do not cancel $\sin x$ as this leads to loss of solutions.

$\sin x(1 - 2 \cos x) = 0$

Either $\sin x = 0 \Rightarrow x = 0, \pi, 2\pi$

or $1 - 2 \cos x = 0 \Rightarrow \cos x = 0.5 \Rightarrow x = 1.047..., 5.235...$

These are the values.

So $x = 0, 1.05$ (2 d.p.)$, \pi, 5.24$ (2 d.p.)$, 2\pi$

3

(a) Given that $\log_a x - \log_a 8 + 2 \log_a 4 = 0$, where a is a positive constant, find x.

(b) Find the value of x, correct to two decimal places, that satisfies the equation
$$3^{2x+1} = 40.$$

(a) $\log_a x - \log_a 8 + 2\log_a 4 = 0$

Simplify, using the log laws.

$\Rightarrow \log_a x - \log_a 16 + \log_a 8 = 0$

$\Rightarrow \log_a\left(\dfrac{16x}{8}\right) = 0$

$\Rightarrow \log_a(2x) = 0$

Remember that $a^0 = 1$ for all values of a.

$\Rightarrow 2x = 1$

$x = 0.5$

(b) $3^{2x+1} = 40$

Take logs to the base 10 of both sides.

$\log_{10}(3^{2x+1}) = \log_{10}40$

$(2x + 1)\log_{10}3 = \log_{10}40$

Use $\log_a(b^c) = c \log_a b$

$2x + 1 = \dfrac{\log_{10}40}{\log_{10}3} = 3.3577...$

$2x = 2.3577...$

$x = 1.18$ (2 d.p.)

Sample questions and model answers *(continued)*

4

The third term of a geometric series is 20 and the fifth term is 5.

Given that the common ratio, r, is negative, find

(a) the value of r

(b) the first term,

(c) the sum to infinity of the series.

> The nth term of a geometric series is ar^{n-1}.

(a) $ar^2 = 20$ (1)

$ar^4 = 5$ (2)

Dividing (2) by (1) gives $r^2 = \frac{1}{4}$

$r < 0 \Rightarrow r = -\frac{1}{2}$.

(b) Substitute into (1) $a(\frac{1}{4}) = 20$

$a = 80$

> The series is convergent, since $|r| < 1$.

(c) $S_\infty = \dfrac{a}{1-r}$

$= \dfrac{80}{1-(-\frac{1}{2})}$

$= 53\frac{1}{3}$

5

(a) Express $\dfrac{5x^2 - 1}{\sqrt{x}}$ in the form $5x^a - x^b$, where a and b are rational numbers to be found.

(b) Hence find the exact value of $\displaystyle\int_1^2 \dfrac{5x^2 - 1}{\sqrt{x}} \, dx$.

(a) $\dfrac{5x^2 - 1}{\sqrt{x}} = \dfrac{5x^2}{x^{1/2}} - \dfrac{1}{x^{1/2}} = 5x^{3/2} - x^{-1/2}$

> Remember that $\sqrt{x} = x^{1/2}$ and work in index form.

So $a = \frac{3}{2}$ and $b = -\frac{1}{2}$.

(b) $\displaystyle\int_1^2 \dfrac{5x^2 - 1}{\sqrt{x}} \, dx = \int_1^2 (5x^{3/2} - x^{-1/2}) \, dx$

$= \left[\dfrac{5x^{5/2}}{\frac{5}{2}} - \dfrac{x^{1/2}}{\frac{1}{2}} \right]_1^2$

> $(2)^{5/2} = (\sqrt{2})^5 = 4\sqrt{2}$

$= \left[2x^{5/2} - 2x^{1/2} \right]_1^2$

> Leave the answer in surd form.

$= 8\sqrt{2} - 2\sqrt{2} - (2 - 2)$

$= 6\sqrt{2}$

Practice examination questions

1 (a) Write down the exact value of $49^{-\frac{1}{2}}$.

   (b) Find the value of k such that $a^k = \dfrac{a \times \sqrt{a^3}}{\sqrt{a}}$.

2 (a) Given that $8^{x-3} = 4^{x+1}$, find the value of x.

   (b) Solve the equation $5x^{-\frac{1}{3}} = x^{\frac{1}{6}}$.

3 The 10th term of an arithmetic progression is 74 and the sum of the first 20 terms is 1510.
Find the first term and the common difference.

4 (a) Find the 8th term of a geometric progression with first term 3 and common ratio 2.

   (b) Find the sum of the first 50 terms of the series $25 + 26.2 + 27.4 + \dots$.

5 (a) Express $3\log_a x - 2\log_a y + \log_a(x+1)$ as a single logarithm.

   (b) Solve the equation $5^x = 100$. Give your answer to 3 d.p.

6 Solve the equation $\cos(2x + 30°) = 0.4$ for $0° \leqslant x \leqslant 360°$.

7 (a) Write the equation $\cos x + 3 \sin x \tan x - 2 = 0$ in the form $a\cos^2 x + b\cos x + c = 0$.

   (b) Find the solutions of $\cos x + 3 \sin x \tan x - 2 = 0$ for $0° \leqslant x \leqslant 360°$.

8 Given that $f(x) = x^2 + \dfrac{1}{x}$, find

   (a) $f'(2)$.

   (b) $f''(-1)$.

Practice examination questions (continued)

9 The diagram shows a region R bounded by the curve $y = 3\sqrt{x}$, the x-axis and the lines $x = 1$ and $x = 4$.

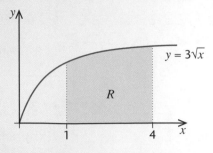

Calculate the area of the region R.

10 The diagram shows a circle of unit radius.

The angle between the radii is θ where

$0 < \theta < \pi$ and the area of the shaded

segment is $\dfrac{\pi}{5}$.

Show that $\theta = \sin\theta + \dfrac{2\pi}{5}$.

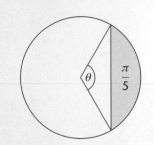

11 The area shaded in the diagram is bounded by the curve $y = \sqrt{x^2 - 3}$, the lines $x = 2$ and $x = 3$ and the x-axis.

Express the shaded area as an integral and estimate its value using the trapezium rule with five intervals. Give your answer to 2 d.p.

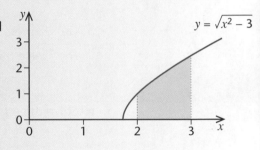

12 (a) Factorise $2x^{\frac{3}{2}} - \dfrac{2x^{\frac{5}{2}}}{5}$.

(b) Find the points at which the graph of $y = 2x^{\frac{3}{2}} - \dfrac{2x^{\frac{5}{2}}}{5}$ meets the x-axis.

(c) Find $\dfrac{dy}{dx}$.

(d) Solve $\dfrac{dy}{dx} = 0$ for $x > 0$. Hence find the coordinates of the turning point of the graph, where x is positive, and determine its nature.

Chapter 3

Mechanics 1

The following topics are covered in this chapter:

- Vectors
- Kinematics
- Statics
- Dynamics
- Projectiles

3.1 Vectors

Modelling

To solve a problem in mechanics, we usually need to make some simplifying assumptions from the outset. For example, to analyse the motion of a projectile we will assume that air resistance is small enough to be ignored and that the acceleration due to gravity is constant. This greatly simplifies the equations used so it is a lot easier to solve the problem.

The set of simplifying assumptions that we apply is referred to as the mathematical **model** of the situation.

In the exam, the questions are worded so that the model is clear. For example, a pulley may be described as light and frictionless; a string may be light and inextensible and a large object may be treated as a particle. However, you may be asked to comment on the extent to which the simplification of the model affects the accuracy of the results.

Vector and scalar quantities

A **scalar** quantity has size (or **magnitude**) but not direction. **Numbers** are scalars and some other important examples are **distance**, **speed**, **mass** and **time**.

A **vector** quantity has both size and **direction**. For example, distance in a specified direction is called **displacement**. Some other important examples are **velocity**, **acceleration**, **force** and **momentum**.

The diagram shows a **directed line segment**. It has size (in this case, length) and direction so it is a vector.

In a textbook, vectors are usually labelled with lower case letters in **bold** print. When hand-written, these letters should be underlined e.g. *a*

The diagram gives a useful way to represent *any* vector quantity and may also be used to represent addition and subtraction of vectors and multiplication of a vector by a scalar.

Addition and subtraction of vectors

This diagram shows three vectors **a**, **b** and **c** such that **c** = **a** + **b**.

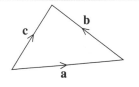

The vectors **a** and **b** follow on from each other and then **c** joins the start of vector **a** to the end of vector **b**.

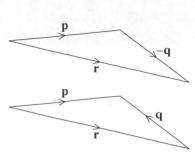

Check that **a** + **b** = **b** + **a**.

The vector **c** is called the **resultant** of **a** and **b**.

On a vector diagram, −**q** has the opposite sense of direction to **q**.

Notice that **p** − **q** is represented as **p** + (−**q**).

This diagram shows the same information in a different way.

Following the route in the opposite direction to **q** is the same as adding −**q** or subtracting **q**.

Scalar multiplication

2**p** is parallel to **p** and has the same sense of direction but is twice as long.

−3**p** is parallel to **p** but has the opposite sense of direction and is three times as long.

Component form

When working in two dimensions, it is often very useful to express a vector in terms of two special vectors **i** and **j**. These are **unit vectors** at right-angles to each other.

A vector **r** written as **r** = a**i** + b**j** is said to have **components** a**i** and b**j**.

A unit vector is a vector of magnitude 1 unit.

Examples

r = **i** + 2**j**

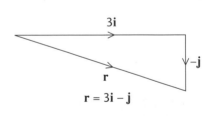

r = 3**i** − **j**

For work involving the Cartesian coordinate system, **i** and **j** are taken to be in the positive directions of the x- and y-axes respectively.

In three dimensions, a third vector **k** is used to represent a unit vector in the positive direction of the z-axis.

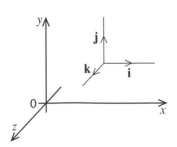

The **position vector** of a point P is the vector $\overrightarrow{OP}$ where O is the origin.

If P has coordinates (a, b) then its position vector is given by **r** = a**i** + b**j**.

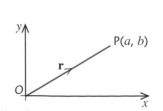

In three dimensions, a point with coordinates (a, b, c) would have position vector **r** = a**i** + b**j** + c**k**.

Resolving a vector

If you know the magnitude and direction of a vector then you can use trigonometry to **resolve** it in terms of **i** and **j** components.

The magnitude of **r** may be written as $|\mathbf{r}|$ or simply as r.

From the diagram, $\mathbf{r} = r \cos \theta\mathbf{i} + r \sin \theta\mathbf{j}$.

Example

A force **f** has magnitude 10 N at an angle of 60° above the horizontal. Express **f** in the form $a\mathbf{i} + b\mathbf{j}$ where **i** and **j** are unit vectors in the horizontal and vertical directions respectively.

In this case, $a = 10 \cos 60° = 5$ and $b = 10 \sin 60° = 5\sqrt{3}$ giving $\mathbf{f} = 5\mathbf{i} + 5\sqrt{3}\mathbf{j}$.

Adding and subtracting vectors in component form

One advantage of expressing vectors in terms of **i** and **j** is that addition and subtraction may be done algebraically.

Example

$\mathbf{p} = 3\mathbf{i} + 2\mathbf{j}$ and $\mathbf{q} = 5\mathbf{i} - \mathbf{j}$. Express the following vectors in terms of **i** and **j**:

(a) $\mathbf{p} + \mathbf{q}$ (b) $\mathbf{p} - \mathbf{q}$ (c) $2\mathbf{p} - 3\mathbf{q}$.

(a) $\mathbf{p} + \mathbf{q} = (3\mathbf{i} + 2\mathbf{j}) + (5\mathbf{i} - \mathbf{j}) = 8\mathbf{i} + \mathbf{j}$

(b) $\mathbf{p} - \mathbf{q} = (3\mathbf{i} + 2\mathbf{j}) - (5\mathbf{i} - \mathbf{j}) = -2\mathbf{i} + 3\mathbf{j}$

(c) $2\mathbf{p} - 3\mathbf{q} = 2(3\mathbf{i} + 2\mathbf{j}) - 3(5\mathbf{i} - \mathbf{j}) = 6\mathbf{i} + 4\mathbf{j} - 15\mathbf{i} + 3\mathbf{j} = -9\mathbf{i} + 7\mathbf{j}$.

Finding the magnitude and direction of a vector

Another advantage of expressing a vector in terms of **i** and **j** components is that it is easy to find its magnitude and direction.

If $\mathbf{r} = a\mathbf{i} + b\mathbf{j}$ then the magnitude of **r** is given by

$$|\mathbf{r}| = \sqrt{a^2 + b^2}.$$

And the direction of r relative to i is given by

$$\theta = \tan^{-1}\left(\frac{b}{a}\right).$$

For example, if $\mathbf{r} = 3\mathbf{i} + 4\mathbf{j}$ then $|\mathbf{r}| = \sqrt{3^2 + 4^2} = 5$ and
$$\theta = \tan^{-1}\left(\tfrac{4}{3}\right) = 53.1°$$

Progress check

1 Write down the position vector **r** of a point with coordinates (5, −2).

2

Given that $|\mathbf{r}| = 12$, express **r** in the form $\mathbf{r} = a\mathbf{i} + b\mathbf{j}$.

3 Find the magnitude of the vector $\mathbf{r} = 5\mathbf{i} - 12\mathbf{j}$ and give its direction relative to **i**.

1 $\mathbf{r} = 5\mathbf{i} - 2\mathbf{j}$ 2 $\mathbf{r} = 6\sqrt{3}\mathbf{i} + 6\mathbf{j}$ 3 $|\mathbf{r}| = 13$, 67.4° clockwise.

3.2 Kinematics

After studying this section you should be able to:

- *apply the constant acceleration formulae for motion in a straight line*
- *draw and interpret graphs of displacement, velocity and acceleration against time*
- *use vectors to analyse motion in two and three dimensions*
- *apply the constant acceleration formulae to projectiles*

LEARNING SUMMARY

Motion in a straight line

You can use these formulae whenever the acceleration of an object is constant.

$v = u + at$	s is the displacement from a fixed position
$s = ut + \frac{1}{2}at^2$	u is the initial velocity
$s = \left(\dfrac{u + v}{2}\right)t$	a is the acceleration
	t is the time that the object has been in motion
$v^2 = u^2 + 2as$	v is the velocity at time t.

Example

An object starts from rest and moves in a straight line with constant acceleration 3 m s^{-2}. Find its velocity after 5 s.

From the given information: $u = 0$

$a = 3$

$t = 5$.

> Don't include units in your working.

> Make a clear statement and include the appropriate units.

Using $v = u + at$

gives $v = 0 + 3 \times 5 = 15$.

The velocity after 5 s is 15 m s^{-1}.

> **KEY POINT**
>
> Motion may take place in either direction along a straight line. One direction is taken to be positive for displacement, velocity and acceleration and the other direction is taken to be negative.

> In some questions the acceleration due to gravity is taken to be 9.8 m s^{-2}.

Example

A stone is thrown vertically upwards with a velocity of 20 m s^{-1}. It has a downward acceleration due to gravity of 10 m s^{-2}.

(a) Find the greatest height reached by the stone.

(b) Find its velocity after 3 s.

(c) Find the height of the stone above the point of projection after 3 s.

(a) From the given information: $u = 20$

$a = -10$.

At the greatest height $v = 0$.

Using $v = u + at$

$0 = 20 - 10t \Rightarrow t = 2$.

Using $s = ut + \frac{1}{2}at^2$

gives $s = 20 \times 2 - 5 \times 4 = 20$

The greatest height reached is 20 m.

(b) Using

$v = u + at$ with $t = 3$

gives

$v = 20 - 10 \times 3 = -10.$

After 3 s the stone is moving downwards at 10 m s^{-1}.

(c) Using

$s = ut + \frac{1}{2}at^2$

gives

$s = 20 \times 3 - 5 \times 3^2 = 15.$

After 3 s the stone is 15 m above the point of projection.

Graphical representation

You need to know the properties of the graphs of distance, displacement, speed, velocity and acceleration against time.

> The gradient of a distance–time graph represents speed.
> The gradient of a displacement–time graph represents velocity.
> The gradient of a velocity–time graph represents acceleration.
>
> The area under a speed–time graph represents the distance travelled.
> The area under a velocity–time graph represents the change of displacement.
> The area under an acceleration–time graph represents change in velocity.

KEY POINT

Example

The diagram represents the progress of a car as it travels between two sets of traffic lights.
Find:

(a) The initial acceleration of the car.

(b) The deceleration of the car as it approaches the second set of lights.

(c) The distance between the traffic lights.

(a) Gradient of OA $= \frac{15}{10} = 1.5$
The initial acceleration of the car is 1.5 m s^{-2}.

(b) Gradient of BC $= -\frac{15}{15} = -1$.
The *acceleration* of the car is -1 m s^{-2} so the *deceleration* is 1 m s^{-2}.

(c) The area of the trapezium OABC is given by $\frac{15}{2}(5 + 30) = 262.5$,
so, the distance between the traffic lights is 262.5 m.

Using vectors

The work of this section may be extended to two and three dimensions using vectors.

Example

A particle has velocity $2\mathbf{i} - 3\mathbf{j}$ m s^{-1} when $t = 0$ and a constant acceleration of $\mathbf{i} + \mathbf{j}$ m s^{-2}. Find the speed of the particle when $t = 5$ and give its direction.

> The constant acceleration formulae work equally well in two or three dimensions.

From the given information: $\quad \mathbf{u} = 2\mathbf{i} - 3\mathbf{j}$

$\qquad\qquad\qquad\qquad\qquad\quad \mathbf{a} = \mathbf{i} + \mathbf{j}$

$\qquad\qquad\qquad\qquad\qquad\quad t = 5.$

Using $\mathbf{v} = \mathbf{u} + \mathbf{a}t$ gives $\qquad \mathbf{v} = 2\mathbf{i} - 3\mathbf{j} + 5(\mathbf{i} + \mathbf{j}) = 7\mathbf{i} + 2\mathbf{j}.$

Speed is the magnitude of the velocity, giving $\quad v = \sqrt{7^2 + 2^2} = 7.28\ldots$.

So the speed of the particle when $t = 5$ is 7.28 m s^{-1} to 2 d.p.

From the diagram, $\tan\theta = \frac{2}{7} \Rightarrow \theta = 15.9°.$

When $t = 5$ the particle is moving at 15.9° anticlockwise to the direction of $\mathbf{i}$.

Projectiles

In the standard model for projectiles, air resistance is ignored and the acceleration due to gravity g m s^{-2} is taken to be constant. According to this model:

- The only force acting on a projectile, once in flight, is its weight.
- The horizontal component of velocity is constant.
- The vertical component of velocity is subject to a constant acceleration g m s^{-2}.

The approach used for many projectile problems is to resolve the initial velocity into horizontal and vertical components and to treat these two parts separately.

Example

A stone is thrown horizontally with a speed of 10 m s^{-1} from the top of a cliff. The cliff is 98 m high and the stone lands in the sea d m from its base. Take $g = 9.8$ m s^{-2} and show that $d = 20\sqrt{5}$.

> You can assume that a cliff is vertical.

Vertically: $\quad u = 0$

$\qquad\qquad\quad a = 9.8$

$\qquad\qquad\quad s = 98.$

> Choosing downwards to be positive.

Using $\qquad s = ut + \frac{1}{2}at^2$

gives $\qquad 98 = 4.9t^2$

$\qquad\qquad \Rightarrow t = \sqrt{20} = 2\sqrt{5}.$

> Even a very simple diagram can help you to visualise the problem.

Horizontally: $\quad d = 10 \times 2\sqrt{5}$

$\qquad\qquad\qquad \Rightarrow d = 20\sqrt{5}.$

The initial velocity of the projectile may be at angle θ to the horizontal.

Example

A ball is struck with velocity 20 m s^{-1} at 40° to the horizontal from a point 1 m above the ground. Find the maximum height reached by the ball. Take $g = 9.8$ m s^{-2}.

Choosing upwards to be positive.

Vertically: $u = 20 \sin 40°$

$a = -9.8.$

At max ht: $v = 0.$

Using $v = u + at$

$0 = 20 \sin 40° - 9.8t$

$\Rightarrow t = 1.3118 \ldots .$

Using $s = ut + \frac{1}{2}at^2.$

At max ht s

$= 20 \sin 40° \times 1.3118 - 4.9 \times 1.3118^2$

$= 8.432 \ldots .$

The maximum height is 9.43 m to 2 d.p.

Progress check

1 A stone is thrown vertically upwards with a speed of 15 m s^{-1}. Take the downward acceleration due to gravity to be 10 m s^{-2} and find:
   (a) The greatest height reached.
   (b) The speed and direction of the stone after 2 s.

2 A particle has position vector $\mathbf{r} = 5t^2\mathbf{i} + t^3\mathbf{j}$ m at time t s.
   Find its velocity and acceleration when $t = 3$.

2   $\mathbf{v} = 30\mathbf{i} + 27\mathbf{j}$, $\mathbf{a} = 10\mathbf{i} + 18\mathbf{j}$
1 (a) 11.25 m   (b) 5 m s^{-1} downwards

3.3 Statics

After studying this section you should be able to:

- resolve a single force into perpendicular components
- resolve a system of forces in a given direction
- find the resultant of a system of forces
- solve problems involving friction
- apply the conditions for equilibrium of coplanar forces in simple cases

LEARNING SUMMARY

Force

Force is a vector quantity that influences the motion of an object. It is measured in newtons (N). For example, the **weight** of an object is the force exerted on it by gravity. An object with a mass of m kg has weight mg N. **Tension**, **reaction** and **friction** are other examples of force and will be met in this section.

Resolving forces

Here are some examples of resolving forces into two components at right-angles to each other.

In each case the forces are represented in magnitude and direction by the sides of the triangle. The force to be resolved is *always* shown on the hypotenuse.

> Notice the directions of the forces shown by the arrows. The diagram represents vector addition of the components.

> **KEY POINT**
>
> These are **vector diagrams** showing the relationship between a force and its components.

> Take care not to show both the force *and* its components on a **force diagram** or you will represent the force TWICE.

In a **force diagram**, you can *replace* a force with a pair of components that are equivalent to it. This will often make it easier to produce the equations necessary to solve a problem.

An important example is that of an object on an **inclined plane**.

The diagram shows an object of weight W on a smooth plane inclined at angle θ to the horizontal. R is the force that the plane exerts on the object. It acts at right-angles to the plane and is called the **normal reaction**.

To analyse the behaviour of the object the weight is usually resolved into components parallel and perpendicular to the plane.

> It's worth remembering these results.

This vector diagram shows the component of weight acting down the plane is $W \sin \theta$ and the component perpendicular to the plane is $W \cos \theta$.

Force diagram

Vector diagram

Example

An object of weight 10 N is held in place on a plane inclined at $30°$ to the horizontal by two forces F N and R N as shown.

Find the values of F and R.

The forces F and R correspond to the components of the weight, parallel and perpendicular to the plane, but act in the opposite directions.

> The forces must balance in each direction.

Resolving parallel to the plane: $F = 10 \sin 30° = 5$.

Resolving perpendicular to the plane: $R = 10 \cos 30° = 5\sqrt{3}$.

Resultant force

The effect of several forces acting at a point is the same as the effect of a single force called the **resultant force**.

You can find the resultant of a set of forces by using vector addition.

Example

Find the magnitude and direction of the resultant **R** N of the forces $(3\mathbf{i} - 2\mathbf{j})$ N, $(\mathbf{i} + 3\mathbf{j})$ N and $(-2\mathbf{i} + 4\mathbf{j})$ N.

The resultant force is given by: $\mathbf{R} = (3\mathbf{i} - 2\mathbf{j}) + (\mathbf{i} + 3\mathbf{j}) + (-2\mathbf{i} + 4\mathbf{j})$
$$= 2\mathbf{i} + 5\mathbf{j}.$$

The magnitude of the resultant is $|\mathbf{R}| = \sqrt{2^2 + 5^2} = \sqrt{29}$.

From the diagram: $\theta = \tan^{-1}(\tfrac{5}{2}) = 68.2°$.

The resultant has magnitude $\sqrt{29}$ N and acts at $68.2°\ \circlearrowleft$ to the direction of **i**.

Equilibrium

A set of forces acting at a point is in **equilibrium** if the resultant force is zero.

It follows that the resolved parts of the forces must balance in any chosen direction.

Example

Find the force **F** such that the forces $(5\mathbf{i} - 3\mathbf{j})$, $(2\mathbf{i} + \mathbf{j})$ and **F** are in equilibrium.

For equilibrium $(5\mathbf{i} - 3\mathbf{j}) + (2\mathbf{i} + \mathbf{j}) + \mathbf{F} = 0$
$$\Rightarrow 7\mathbf{i} - 2\mathbf{j} + \mathbf{F} = 0$$
$$\Rightarrow \mathbf{F} = -7\mathbf{i} + 2\mathbf{j}.$$

Friction

Whenever two rough surfaces are in contact, the tendency of either surface to move relative to the other is opposed by the **force of friction**.

The diagram shows an object of weight W N resting on a rough horizontal surface. The object is pushed from one side by a force P N and friction responds with force F N in the opposite direction.

For small values of P, no movement takes place and $F = P$.

If P increases then F increases to maintain equilibrium until F reaches a maximum value. At this point the object is in **limiting equilibrium** and is on the point of slipping.

If P is now increased again then F will remain the same. The equilibrium will be broken and the object will move.

The maximum value of F depends on:

- The magnitude of the normal reaction R.
- The roughness of the two surfaces measured by the value μ. This value is known as the **coefficient of friction**.

In general: $F \leqslant \mu R$.
For limiting equilibrium: $F = \mu R$.

Example
An object of mass 8 kg rests on a rough horizontal surface. The coefficient of friction is 0.3 and a horizontal force P N acts on the object which is about to slide. Take $g = 10 \text{ m s}^{-2}$ and find the value of P.

> Remember that the weight of the object is given by mg.

Resolving vertically gives $R = 80$.

Friction is limiting so $F = \mu R = 0.3 \times 80 = 24$.

Resolving horizontally: $P = F \Rightarrow P = 24$.

Example
In the diagram, the object of mass m kg is on the point of slipping down the plane. The coefficient of friction between the object and the plane is μ. Show that $\mu = \tan \theta$.

Resolving perpendicular to the plane $R = mg \cos \theta$.
Resolving parallel to the plane $F = mg \sin \theta$.
Friction is limiting so $F = \mu R$.

This gives $mg \sin \theta = \mu mg \cos \theta \Rightarrow \mu = \dfrac{\sin \theta}{\cos \theta} = \tan \theta$.

Progress check

1 An object of mass 8 kg rests on a plane inclined at 40° to the horizontal. Find the components of its weight parallel and perpendicular to the plane. Take $g = 9.81$ m s^{-2}.

2 Find the magnitude and direction of the resultant **R** N of the forces $(2\mathbf{i} + 3\mathbf{j})$ N, $(5\mathbf{i} - 4\mathbf{j})$ N and $(-3\mathbf{i} - 7\mathbf{j})$ N.

3 An object of weight 50 N rests on a rough horizontal surface. A horizontal force of 20 N is applied to the object so that it is on the point of slipping. Find the value of the coefficient of friction.

3 0.4
2 8.94 N at 63.4° ↻ to **i**
1 parallel: 50.4 N, perp: 60.1 N

3.4 Dynamics

After studying this section you should be able to:

- understand and apply Newton's laws of motion in two or three dimensions
- analyse the motion of connected particles
- understand and apply the principle of conservation of momentum

LEARNING SUMMARY

Newton's laws of motion

Newton's laws of motion provide us with a clear set of rules that can be used to analyse the effect of forces within a system.

The laws may be stated as:
1. Every particle continues in a state of rest or uniform motion, in a straight line, unless acted upon by an external force.
2. The resultant force acting on a particle is equal to its rate of change of momentum. *This law is most often applied in the form $F = ma$.*
3. Every force has an equal and opposite reaction.

The formula $F = ma$

In the formula $F = ma$, F N stands for the resultant force acting on a particle, m kg is the mass of the particle and a m s^{-2} is the acceleration produced.

It is important to use the correct units.

Example
A particle of mass 2 kg rests on a smooth horizontal plane.
Horizontal forces of 15 N and 4 N act on the particle in opposite directions.
Find the acceleration of the particle.

Using $F = ma$
$$15 - 4 = 2a$$
$$\Rightarrow \qquad a = 5.5$$

15 N → ● ← 4 N
2 kg

The acceleration of the particle is 5.5 m s^{-2}.

The formula applies equally well in two or three dimensions using vectors.

Example
Forces $(7\mathbf{i} - 2\mathbf{j} + \mathbf{k})$ N and $(3\mathbf{i} + \mathbf{j} - 5\mathbf{k})$ N act on a particle of mass 10 kg.
Find the acceleration produced.

The resultant force is the vector sum of the given forces, so using $\mathbf{F} = m\mathbf{a}$ gives

$$(7\mathbf{i} - 2\mathbf{j} + \mathbf{k}) + (3\mathbf{i} + \mathbf{j} - 5\mathbf{k}) = 10\mathbf{a}$$
$$\Rightarrow \qquad 10\mathbf{i} - \mathbf{j} - 4\mathbf{k} = 10\mathbf{a}$$
$$\Rightarrow \qquad \mathbf{a} = \mathbf{i} - 0.1\mathbf{j} - 0.4\mathbf{k}$$

The acceleration of the particle is $\mathbf{i} - 0.1\mathbf{j} - 0.4\mathbf{k}$ m s^{-2}.

The formula applies even when the force is variable.

Example

A force $(2t\mathbf{i} - 5\mathbf{j})$ N acts on a particle of mass 0.5 kg at time t seconds. Find the velocity of the particle at time t given that, initially, the velocity is $-3\mathbf{i} + 7\mathbf{j}$ m s^{-1}.

Using $\mathbf{F} = m\mathbf{a}$ gives $\quad 2t\mathbf{i} - 5\mathbf{j} = 0.5\mathbf{a}$

$$\Rightarrow \qquad \mathbf{a} = 4t\mathbf{i} - 10\mathbf{j}$$

$$\mathbf{v} = \int 4t\mathbf{i} - 10\mathbf{j} \, dt$$

$$\Rightarrow \qquad \mathbf{v} = 2t^2\mathbf{i} - 10t\mathbf{j} + \mathbf{c}.$$

When $t = 0$, $\mathbf{v} = -3\mathbf{i} + 7\mathbf{j} \Rightarrow \mathbf{c} = -3\mathbf{i} + 7\mathbf{j}$.

This gives $\qquad\qquad\qquad \mathbf{v} = 2t^2\mathbf{i} - 10t\mathbf{j} - 3\mathbf{i} + 7\mathbf{j}$

$$\mathbf{v} = (2t^2 - 3)\mathbf{i} + (7 - 10t)\mathbf{j}.$$

The velocity of the particle at time t is $(2t^2 - 3)\mathbf{i} + (7 - 10t)\mathbf{j}$ m s^{-1}.

Connected particles

In a typical problem, two particles are connected by a **light inextensible string**.
Since the string is light, there is no need to consider its mass.
Since it is inextensible, both particles will have the same speed and accelerate at the same rate *while the string is taut*.

By Newton's third law, the tension in the string acts equally on both particles but in opposite directions.

To solve problems involving connected particles you need to:

* Draw a diagram.

* Identify and label the forces acting on each particle.

* Write the **equation of motion** for each particle, i.e. apply $F = ma$ for each one.

* Solve the simultaneous equations produced.

Example

Two particles P and Q are connected by a light inextensible string. The particles are at rest on a smooth horizontal surface and the string is taut. A force of 10 N is applied to particle Q in the direction PQ. P has mass 2 kg and Q has mass 3 kg. Find the tension in the string and the acceleration of the system.

Note the use of the double headed arrow ➤➤ to represent acceleration.

Using $F = ma$ For particle P $T = 2a$ [1]

                  For particle Q $10 - T = 3a$ [2]

[1] + [2] gives $10 = 5a \Rightarrow a = 2$

Substituting for a in [1] gives $T = 4$.

The tension in the string is 4 N and the acceleration of the system is 2 m s^{-2}.

Example

Particles of mass 3 kg and 5 kg are connected by a light inextensible string. The string passes over a light, frictionless pulley and hangs vertically.

The particles are released from rest with the string taut.

Find the acceleration of the system and the tension in the string.

Take $g = 9.8$ m s^{-2}

Draw the diagram.

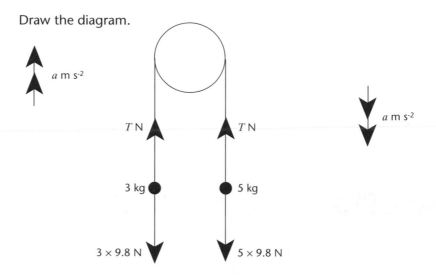

	KEY POINT
An important consequence of the model is that the tension in the string has the same value on both sides of the pulley.	

Using $F = ma$ separately for each particle gives two simultaneous equations.

For the 3 kg particle: $T - 3 \times 9.8 = 3a$ (1)

For the 5 kg particle: $5 \times 9.8 - T = 5a$ (2)

(1) + (2) gives $5 \times 9.8 - 3 \times 9.8 = 8a$

So $$a = \frac{2 \times 9.8}{8} = 2.45$$

Substituting for a in (1) gives $T = 3(2.45 + 9.8) = 36.75$

The acceleration of the system is 2.45 m s^{-2}

The tension in the string is 36.75 N.

Momentum

The **momentum** of a particle is a vector quantity given by the product of its mass and its velocity, i.e. momentum = $m\mathbf{v}$ where m kg is the mass of a particle and $\mathbf{v}$ m s^{-1} is its velocity.

The principle of conservation of momentum states that when two particles collide:

the total momentum before impact = the total momentum after impact.

> This kind of diagram gives a useful way to represent the information.

Before impact

After impact

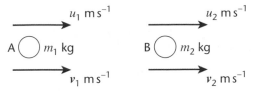

> The positive direction for velocity is shown from left to right.

In the diagram,
$u_1 > u_2$ so that the particles collide.

Conservation of momentum gives $m_1 u_1 + m_2 u_2 = m_1 v_1 + m_2 v_2$.

Example

A particle of mass 5 kg moving with speed 4 m s^{-1} hits a particle of mass 2 kg moving in the opposite direction with speed 3 m s^{-1}. After the impact the two particles move together with the same speed v m s^{-1}. Find the value of v.

> Unknown velocities are shown in the positive direction.

Before impact

After impact

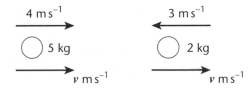

Conservation of momentum gives $5 \times 4 - 2 \times 3 = 5v + 2v$

$\Rightarrow \qquad 7v = 14$

$\Rightarrow \qquad v = 2.$

Progress check

1 A particle of mass 8 kg is acted upon by the forces $(11\mathbf{i} - 3\mathbf{j})$ N, $(14\mathbf{i} + \mathbf{j})$ N and $(-\mathbf{i} + 10\mathbf{j})$ N. Find the acceleration of the particle.

2 A particle moves along the x-axis such that its displacement x m from O at time t s is given by $x = t^3 - 15$ for $0 \leqslant t \leqslant 5$.

   (a) Find the acceleration of the particle at time t s.

   (b) Given that the particle has mass 5 kg find the resultant force acting on it when $t = 4$.

3 Two particles A and B are connected by a light inextensible string. The particles are at rest on a smooth horizontal surface and the string is taut. A force of 12 N is applied to particle B in the direction AB. A has mass 4 kg and B has mass 2 kg. Find the tension in the string and the acceleration of the system.

4 A particle of mass 8 kg moving with speed 2 m s^{-1} hits a stationary particle of mass 2 kg.
   After the impact both particles move in the same direction with speed v m s^{-1}.
   Find the value of v.

1 $3\mathbf{i} + \mathbf{j}$ m s^{-2}.   2 (a) $6t$ m s^{-2} (b) 120 N.   3 Tension = 8 N, acceleration = 2 m s^{-2}.   4 $v = 1.6$

3.5 Projectiles

After studying this section you should be able to:

- model the motion of a projectile moving under constant acceleration
- understand the limitations of the model
- solve problems by considering, separately, the vertical and horizontal components of velocity

LEARNING SUMMARY

The model

In the usual **particle** model for projectiles, air resistance is ignored and the only force taken to act on the particle is its weight, which is constant.
According to this model:

> Check the value of g used on your exam paper. Make sure that you use the given value or you may lose marks.

KEY POINT

- The horizontal component of velocity remains constant throughout the motion.
- The vertical component of velocity is subject to a constant downward acceleration of g m s^{-2}.

The accuracy of the model, in predicting the motion of a projectile, depends on the extent to which the initial assumptions are satisfied.

For example, the model will provide accurate information about a projectile such as a stone or a dart over a short distance, but will give poor results for light objects that are affected more by air resistance.

> Be careful about using formulae for time of flight, greatest height, range etc. without justification or, again, marks may be lost.

When using the model, a first step is usually to resolve the velocity of a projectile into horizontal and vertical components. The two parts are then treated separately.

A particle moving with speed v at angle θ to the horizontal has vertical component $v \sin \theta$ and horizontal component $v \cos \theta$.

> The velocity is usually shown on a diagram as in Figure 1 but is used in equations as in Figure 2.

If you know the horizontal and vertical components of a velocity then you can combine them to find its magnitude (speed) and direction.

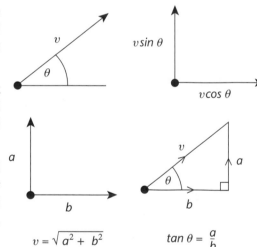

Figure 1   Figure 2

$$v = \sqrt{a^2 + b^2} \qquad \tan \theta = \frac{a}{b}$$

Upwards is taken to be positive so the acceleration is shown as negative.

Example

A stone is thrown with speed 15 m s^{-1} at an angle of 60° above the horizontal. Find its speed and direction after 1 second. Take $g = 9.8$ m s^{-2}.

Vertically: using $v = u + at$

$$v_a = 15 \sin 60° - 9.8$$

$$= 3.19038 \ldots$$

Horizontally: $v_b = 15 \cos 60°$

$$= 7.5$$

$$v = \sqrt{7.5^2 + 3.19038^2} = 8.15 \text{ to 3 s.f.}$$

$$\tan \theta = \frac{3.19038}{7.5} \Rightarrow \theta = 23.0° \text{ to 1 d.p.}$$

After 1 second the stone has speed 8.15 m s^{-1} at 23° above the horizontal.

Progress check

1 A small object is projected from ground level with speed 30 ms^{-1} at 40° above the horizontal. Take $g = 10$ ms^{-2}.

(a) Find the speed and direction of the object after 1.5 seconds.

(b) The object just clears a tree 25 m from the point of projection. How high is the tree?

1 (a) 23.4 m s^{-1}, 10.6° above horiz (b) 15.2 m

Sample questions and model answers

1

A particle moving in a straight line passes through a fixed point O when $t = 0$. Its velocity at time t seconds is given by $v = 12 - 3t^2$ m s^{-1}.

(a) Find the acceleration of the particle when $t = 4$.

(b) Find the distance of the particle from O:

　　(i)　when it comes to rest
　　(ii)　when $t = 5$.

(a) $v = 12 - 3t^2 \implies a = \dfrac{dv}{dt} = -6t$

　　$t = 4 \implies a = -6 \times 4 = -24.$

The acceleration of the particle when $t = 4$ is -24 m s^{-2}.

(b) (i)　The particle comes to rest when $v = 0$.

　　　$v = 0 \implies 12 - 3t^2 = 0$
　　　　　　$\implies t = 2$ since $t > 0$.

　　The displacement of the particle from O at time t is given by

$$x = \int v\, dt + c = \int 12 - 3t^2\, dt + c$$

　　$\implies x = 12t - t^3 + c.$

　　When $t = 0$, $x = 0$

　　giving　　　　　　$0 = 0 + c \implies c = 0.$

　　So　　　　　　　$x = 12t - t^3.$

　　When $t = 2$　　$x = 12 \times 2 - 2^3 = 16.$

　　The particle is 16 m from O when it comes to rest.

(b) (ii)　When $t = 5$　$x = 12 \times 5 - 5^3 = -65.$

Distance is taken to be positive.

　　　The particle is 65 m from O when $t = 5$.

Sample questions and model answers (continued)

2

Two particles are connected by a light inextensible string passing over a light frictionless pulley. Particle A has mass 5 kg and lies on a smooth horizontal table. Particle B has mass 4.5 kg and hangs freely. Take $g = 10\ \text{m s}^{-2}$.

(a) Write down the equation of motion for each particle.

(b) Find the acceleration of the system.

(c) Find the tension in the string.

(a)

A clearly labelled diagram is an essential first step.

Both particles have the same acceleration.

The tension is the same at both ends of the string.

The diagram shows the forces acting on each particle in the direction of motion.

Use $F = ma$ for each particle in turn.

The equations of motion are:

for particle A $T = 5a$ [1]

for particle B $45 - T = 4.5a$ [2]

Solve the simultaneous equations and interpret the results.

(b) [1] + [2] gives $45 = 9.5a$

$\Rightarrow a = 4.74$ to 3 s.f.

The acceleration of the system is $4.74\ \text{ms}^{-2}$ to 3 s.f.

(c) Substitution for a in [1] gives $T = 23.7$ to 3 s.f.

It is a good idea to state the degree of accuracy given in your answer.

The tension in the string is 23.7 N to 3 s.f.

Sample questions and model answers *(continued)*

3

A stone is projected from ground level with speed 16 ms^{-1} at an angle of 60° to the horizontal.

(a) Find the maximum height reached by the stone.

(b) Calculate the length of time for which the height of the stone is at least 3 m.

A diagram helps to clarify the situation.

16 m s^{-1}

3 m

60°

(a)   Initial vertical component of velocity = 16sin60°

                                   = $8\sqrt{3}$ ms^{-1}

Using $v^2 = u^2 + 2as$ in the vertical direction

At maximum height h m: $v = 0$, $u^2 = (8\sqrt{3})^2 = 192$, $a = -9.8$, $s = h$

Giving $0 = 192 - 2 \times 9.8h$

               $h = 192 \div 19.6$

                 $= 9.7959...$

Maximum height = 9.80 m (3 s.f.)

(b)   Using $s = ut + \dfrac{1}{2}at^2$ with $s = 3$, $u = 8\sqrt{3}$, $a = -9.8$

$3 = 8\sqrt{3}t - 4.9t^2$

So, $4.9t^2 - 8\sqrt{3}t + 3 = 0$

giving $t = \dfrac{8\sqrt{3} \pm \sqrt{192 - 12 \times 4.9}}{2 \times 4.9}$

$t = 2.59159...$ or $t = 0.23624...$

The larger value is the time when the stone is 3 m high on the way down and the smaller value is when the stone is 3 m high on the way up.

Required time = 2.59159... − 0.23624...

                  = 2.36 seconds (3 s.f.)

Practice examination questions

1 A particle moving in a straight line passes through a point O with velocity 10 m s^{-1} when $t = 0$. Given that the acceleration of the particle is -4 m s^{-2}, find:

(a) The velocity of the particle when $t = 5$.

(b) The distance that the particle travels between $t = 0$ and $t = 4$.

2

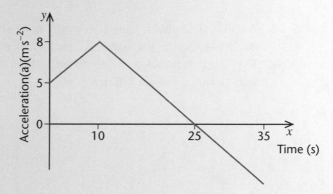

The diagram is an acceleration–time graph for the motion of a particle during a period of 35 s. The particle moves in a straight line and its initial velocity is 20 m s^{-1}.

(a) State the time at which the velocity reaches its maximum value.

(b) Calculate the maximum value of the velocity of the particle during this period.

3 A particle passes through a point P with position vector $(-5\mathbf{i} + 11\mathbf{j} + \mathbf{k})$ with velocity $(2\mathbf{i} + \mathbf{j})$ m s^{-1} when $t = 0$. The particle moves with constant acceleration $(\mathbf{i} + 3\mathbf{k})$ m s^{-2} and reaches the point Q when $t = 4$.

(a) Find $\overrightarrow{PQ}$.

(b) Find the position vector of Q.

(c) Find the average speed of the particle between P and Q.

Practice examination questions *(continued)*

4 A particle is projected from a point on horizontal ground with speed v at angle θ to the horizontal.

(a) Find an expression for the maximum height reached during the subsequent motion.

(b) Find an expression for the horizontal distance travelled before it hits the ground.

5 A ball is kicked with speed 15 m s^{-1} at an angle of 30° to the horizontal from a point on level ground. The ball just clears a fence 2 m high.
Find the maximum possible distance between the fence and the point where the ball was kicked. Take $g = 9.8$ m s^{-2}.

6

The diagram shows an object of mass 12 kg resting on a rough horizontal surface.

A force of 50 N acts on the object at an angle of 40° to the horizontal.
Take $g = 10$ m s^{-2}.

(a) Find the normal reaction between the horizontal surface and the object.

(b) Given that the object is in limiting equilibrium, find the coefficient of friction between the object and the horizontal surface.

7 The diagram shows a mass of 6 kg acted upon by a horizontal force of P N.

The coefficient of friction between the object and the plane is 0.3.

Find the minimum value of P required to maintain the object in equilibrium.

Practice examination questions *(continued)*

8

Particles A and B of mass 2 kg and 3 kg respectively are connected by a light inextensible string passing over a smooth frictionless pulley. The system is released from rest with both particles 1 m above the ground. Find:

(a) The acceleration of the system while the string remains taut.

(b) The tension in the string.

(c) The time taken for particle B to reach the ground.

(d) The speed with which particle B hits the ground.

(e) The greatest height reached by particle A.

(f) What difference would it make to your initial equations if the pulley was not light and frictionless?

9

A particle P of mass $2m$ moving with speed $5u$ strikes particle Q of mass m, which is at rest.
After the collision both particles move in the same direction but the speed of particle Q is twice the speed of particle P.

Find the speed of particle P after the collision.

10 A ball is struck with speed 25 m s^{-1} from a point 50 cm above the ground. When it has travelled 20 m horizontally, the ball just clears a fence that is 3 m high.

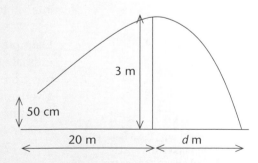

(a) Show that a possible angle of projection of the ball is 16.5°.

(b) Find the distance d m, between the fence and the point where the ball strikes the ground for this angle of projection.

Chapter 4

Statistics 1

The following topics are covered in this chapter:

- Numerical measures
- Probability
- Binomial distribution

- The normal distribution
- Correlation and regression
- Estimation and sampling

4.1 Numerical measures

After studying this section you should be able to:

- calculate averages
- find the variance, standard deviation, range and interquartile range of data
- understand the concepts of outliers

LEARNING SUMMARY

Measures of central location

An average is a value that is taken to be representative of a data set. The three forms of average that you need are the mean, median and mode. These are often referred to as **measures of central location**.

The **mean** $\bar{x}$ of the values $x_1, x_2, x_3, \ldots, x_n$ is given by $\bar{x} = \dfrac{\sum x_i}{n}$. If each x_i occurs with frequency f_i then the mean of the **frequency distribution** is given by $\bar{x} = \dfrac{\sum f_i x_i}{\sum f_i}$.

Example

Find the mean of these results obtained by throwing a dice.

score	1	2	3	4	5	6
frequency	18	17	23	20	24	18

> One problem with the mean is that it may be unduly influenced by a small number of extreme values, known as **outliers**.

$$\bar{x} = \frac{18 \times 1 + 17 \times 2 + 23 \times 3 + 20 \times 4 + 24 \times 5 + 18 \times 6}{18 + 17 + 23 + 20 + 24 + 18} = \frac{429}{120} = 3.575$$

The **median** is the middle value of the data when it is arranged in order of size. If there is an even number of values then the median is the mean of the middle pair. In the example above, the 60th and 61st values are both 4 so the median is 4.

The **mode** is the value that occurs with the highest frequency. In the example above the mode is 5.

You can *estimate* the mean of **grouped data** by using the midpoint of each class interval to represent the class.

Example

Estimate the mean value of h from the figures given in the table.
An estimate of the mean is given by

$$\bar{x} = \frac{755}{72} = 10.486\ldots$$

$$10.5 = \text{to 1 d.p.}$$

> Once data have been grouped, the exact values are not available and so it is only possible to *estimate* the mean.

Interval	frequency (f_i)	midpoint (x_i)	$f_i \times x_i$
$0 < h \leqslant 5$	8	2.5	20
$5 < h \leqslant 10$	24	7.5	180
$10 < h \leqslant 15$	29	12.5	362.5
$15 < h \leqslant 20$	11	17.5	192.5
Totals	72		755

Measures of dispersion

An average alone gives no indication of how widely dispersed the data values are. The simplest measure of dispersion, or spread, is the **range**.

The range of a data set is the difference between largest value and the smallest value.

A slightly more refined approach is to measure the spread of the 'middle half' of the data. The data is put into order and the values Q1, Q2 and Q3 are found that divide the data into quarters.

The diagram illustrates the situation for a data set of 11 values.

Q2 is the **median** of the data.

Q1, the **lower quartile**, is the median of the lower half of the data.

Q3, the **upper quartile**, is the median of the upper half of the data.

The **interquartile range** is then Q3 – Q1.

For n **population** values, the **standard deviation of the population,** σ is given by

Find $x\sigma n$ on your calculator.

$$\sigma = \sqrt{\frac{(x_i - \mu)^2}{n}} = \sqrt{\frac{x_i^2}{n} - \mu^2} \quad \text{where } \mu \text{ is the population mean}$$

For n **sample** values, the **standard deviation of the sample,** s is given by

Find $x\sigma n\text{-}1$ on your calculator.

$$s = \sqrt{\frac{(x_i - \bar{x})^2}{n - 1}} \quad \text{where } \bar{x} \text{ is sample mean}$$

Standard deviation $= \sqrt{\text{variance}}$

The **variance** is the square of the standard deviation.

For grouped data, take the mid-interval value of x.

Progress check

1 For the population 8, 14, 20, 1, 6, 6, 13, 12, 4,
   find   (a) the mean   (b) the standard deviation   (c) the median
               (d) the interquartile range.

2 Find the median, mean and standard deviation of this set of sample data:

x	0	1	2	3
f	4	6	7	3

3

Mass (g)	$20 \leqslant m < 40$	$40 \leqslant m < 50$	$50 \leqslant m < 55$	$55 \leqslant m < 60$	$60 \leqslant m < 90$
Frequency	10	12	10	7	9

For the above population data estimate the mean and the standard deviation.

3 50.89, 14.68
2 1.5, 1.45, 0.999
1 (a) 9.33   (b) 5.56   (c) 8   (d) 8.5

4.2 Probability

After studying this section you should be able to:

- *use Venn diagrams to represent and interpret combined events*
- *use set notation to express and apply probability laws*
- *use tree diagrams to represent a problem*

LEARNING SUMMARY

Venn diagrams and probability laws

Venn diagrams provide a useful way to represent information about **sets** of objects.

The reason for mentioning them here is that the diagrams, and the notation used to express results, have a direct interpretation and application in probability theory.

In each diagram, the rectangle represents the set S of all objects under consideration.

A ∪ B is read as A union B.

The circles represent particular sets A and B of objects within the set S.

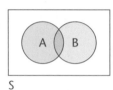

A ∪ B represents the set of all objects that belong to *either* A *or* B *or both*.

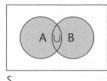

A ∩ B is read as A intersection B.

A ∩ B represents the set of those objects that belong to *both* A *and* B.

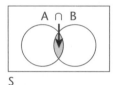

A′ represents the set of objects in S that do not belong to A.

To make use of these ideas in probability theory:

- The objects are interpreted as **outcomes** for a particular situation.
- The set S is the **sample space** of all possible outcomes for the situation.
- The sets A and B are **events** defined by a particular choice of outcomes.
- P(A) for **example**, represents the probability that the event A occurs.
- P(A ∪ B) represents the probability that *either* A *or* B *or both* occurs.
- P(A ∩ B) represents the probability that *both* A *and* B occur.
- P(A′) represents the probability that A does *not* occur.

The **addition rule** states that

P(A ∪ B) = P(A) + P(B) − P(A ∩ B).

A and B are described as **mutually exclusive events** if they have no outcomes in common. These are events that cannot both occur at the same time.

In this case P(A ∩ B) = 0 and the addition law becomes

P(A ∪ B) = P(A) + P(B).   P(A ∩ B) = 0

The events A and A′ are mutually exclusive for any event A.
It follows that P(A ∪ A′) = P(A) + P(A′).
Since one of the events A or A′ must occur, P(A ∪ A′) = 1 so P(A) + P(A′) = 1.

This is usually written as $\quad$ P(A') = 1 – P(A).

The **multiplication rule** states that:

$$P(A \cap B) = P(A \mid B) \times P(B),$$

where P(A|B) means the probability that A occurs *given that* B has *already* occurred.

P(A | B) is described as a **conditional probability**, i.e. it represents the probability of A *conditional* upon B having occurred. Re-arranging the multiplication rule gives

$$P(A \mid B) = \frac{P(A \cap B)}{P(B)}.$$

If A and B are **independent events** then the probability that either event occurs is not affected by whether the other event has already occurred.

In this case, P(A | B) = P(A) and the multiplication rule becomes

$$P(A \cap B) = P(A) \times P(B).$$

Example
The events A and B are independent. P(A) = 0.3 and P(B) = 0.6
Find (a) P(A ∪ B) (b) P(A' ∩ B).

(a) P(A ∪ B) = P(A) + P(B) – P(A ∩ B).

Since A and B are independent, P(A ∩ B) = P(A) × P(B).

> In some problems you need to define the events first before you can apply the probability rules.

So P(A ∪ B) = 0.3 + 0.6 – 0.3 × 0.6 = 0.72

(b) A and B are independent ⇒ A' and B are independent.

So P(A' ∩ B) = P(A') × P(B) = (1 – 0.3) × 0.6 = 0.42

A **tree diagram** is a useful way to represent the probabilities of combined events. Each path through the diagram corresponds to a particular sequence of events and the multiplication rule is used to find its probability. When more than one path satisfies the conditions of a problem, these probabilities are added.

> This is true whether or not the events are independent.

Example
The probability that Chris gets grade A for Maths is 0.6 and the corresponding probability for English is 0.7. The events are independent.

(a) Calculate the probability that Chris gets just one A in the two subjects.
(b) Given that Chris gets just one grade A, find the probability that it is for Maths.

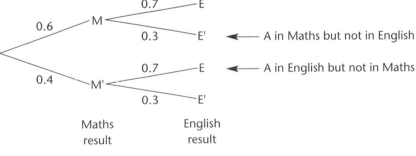

> Show the information on a tree diagram.

0.7 ⟶ E
0.6 / M
0.3 ⟶ E' ⟵ A in Maths but not in English

0.7 ⟶ E ⟵ A in English but not in Maths
0.4 \ M'
0.3 ⟶ E'

Maths result   English result

> Multiply the probabilities along the branches.

(a) P(just one A) = 0.6 × 0.3 + 0.4 × 0.7
$$= 0.18 + 0.28$$
$$= 0.46$$

85

Use conditional probability.

(b) P(A in Maths | just one A) = $\dfrac{\text{P(A in Maths and just one A)}}{\text{P(just one A)}}$

$= \dfrac{0.18}{0.46}$

$= 0.39$ (2 d.p.)

Progress check

1   The probability that the event B occurs is 0.7. The probability that events A and B both occur is 0.4. What is the probability that A occurs given that B has already occurred?

2   It is given that P(A) = 0.4, P(B) = 0.7 and P(A ∩ B) = 0.2.
    Find   (a) P(A ∪ B)   (b) P(B | A)   (c) P(A | B).

3   A bag contains 3 red counters and 7 green counters. Two counters are taken at random from the bag, without replacement.
    (a) By drawing a tree diagram, or otherwise, find the probability that the counters are different colours.
    (b) Given that the counters are different colours, find the probability that the first counter picked is red.

4   The table shows the number of pupils in the lower school.

	Year 7	Year 8	Year 9	Totals
Boy	75	65	55	195
Girl	45	50	40	135
Totals	120	115	95	330

Find the probability that a randomly selected pupil from the lower school:
(a) (i) is a boy
    (ii) is a boy, given that the pupil is from Year 7
    (iii) is from Year 8, given that the pupil is a girl.
(b) Three pupils are selected at random without replacement. Find the probability that:
    (i) all three are from Year 7
    (ii) two boys and one girl are selected
    (iii) a pupil from each year group is selected.

1   4/7

2   (a) 0.9   (b) 0.5   (c) $\frac{2}{7}$

3   (a) $\frac{7}{15}$ (b) $\frac{1}{2}$

4   (a) (i) P(B) = $\frac{195}{330}$

    (ii) P(B|Year 7) = $\frac{75}{120}$

    (iii) P(Year 8|girl) = $\frac{50}{135}$

(b) (i) P(all are Year 7) = $\frac{120}{330} \times \frac{119}{329} \times \frac{118}{328} = 0.0473$ (3 s.f.)

    (ii) P(two boys and one girl) = $3 \times \frac{195}{330} \times \frac{194}{329} \times \frac{135}{328} = 0.430$ (3 s.f.)

    (iii) P(pupil from each year group) = $6 \times$ P(7, 8, 9) = $6 \times \frac{120}{330} \times \frac{115}{329} \times \frac{95}{328} = 0.221$ (3 s.f.)

4.3 Binomial distribution

After studying this section you should be able to:

- *understand the concept of a discrete random variable*
- *understand the binomial distribution*
- *calculate the mean and variance of a binomial distribution*

Discrete random variables

The score obtained when a dice is thrown may be thought of as a **random variable**. Since only specific values may be obtained, this is an example of a **discrete random variable**. In the usual notation:

- A capital letter such as X is used as a label for the random variable.
- A lower case letter such as x is used to represent a particular value of X.
- The probability that X takes the value x is written as $P(X = x)$ or just $p(x)$ and this is known as the **probability function**.

> This is a useful result.

- If X is a discrete random variable with probability function $P(X = x)$ then $\sum P(X = x) = 1$.

The **probability distribution** of X is given by the set of all possible values of x together with the values of $P(X = x)$. This is usually shown in a table.

For example, the probability distribution of the scores shown on a dice may be given as:

> This is an example of a discrete *uniform* distribution.

x:	1	2	3	4	5	6
$p(x)$:	$\frac{1}{6}$	$\frac{1}{6}$	$\frac{1}{6}$	$\frac{1}{6}$	$\frac{1}{6}$	$\frac{1}{6}$

The **cumulative distribution function** is given by $F(x_0) = P(X \leqslant x_0) = \sum_{x \leqslant x_0} p(x)$.

This represents the probability that the random variable X takes a value less than or equal to x_o and is given by the sum of the probabilities up to and including that point.

The binomial distribution

Suppose that **n independent trials** are carried out, each with a **constant probability p of success**.

*n and p are the **parameters** of the distribution.*

If X is the number of successes in n trials, then X follows a **binomial distribution** and this is written $X \sim B(n, p)$.

Use nCr key on the calculator for $\binom{n}{x}$.

Probability $\qquad\qquad P(X = x) = \binom{n}{x} p^x (1 - p)^{n-x} \qquad\qquad x = 0, 1, 2, ..., n$

Mean $\qquad\qquad \mu = np$

Variance $\qquad\qquad \sigma^2 = np(1 - p)$

$\sigma = \sqrt{\text{variance}}$

Standard deviation $\qquad \sigma = \sqrt{np(1 - p)}$

For some values of n and p you can use **cumulative Binomial probability tables** to find $P(X \leqslant x)$ and from this you can find other probabilities, for example:

$P(X < 3) = P(X \leqslant 2)$	**0 1 2 3** ...
$P(X > 3) = 1 - P(X \leqslant 3)$	0 1 2 3 4 **5 6 7** ...
$P(X \geqslant 3) = 1 - P(X \leqslant 2)$	0 1 2 **3 4 5** ...
$P(X = 4) = P(X \leqslant 4) - P(X \leqslant 3)$	0 1 2 3 **4** 5 ...

Example

$X \sim B(10, 0.4)$.

Determine: (a) $P(X > 3)$ (b) $P(3 < X < 7)$ (c) the mean of X

You can use tables when $n = 10$, $p = 0.4$

(a) $P(X > 5) = 1 - P(X \leqslant 5) = 1 - 0.8338 = 0.1662$

(b) $P(3 < X < 7) = P(X \leqslant 6) - P(X \leqslant 3) = 0.9452 - 0.3823 = 0.5629$

(c) Mean $= np = 10 \times 0.4 = 4$

Example

$X \sim B(10, 0.44)$.

$p = 0.44$ is not in the binomial tables so you have to use the formula.

Determine: (a) $P(X = 7)$ (b) the standard deviation of X.

(a) $P(X = 7) = \binom{10}{7} 0.44^7 (1 - 0.44)^3 = 0.06728...$

(b) Standard deviation $= \sigma = \sqrt{10 \times 0.44(1 - 0.44)} = 1.569...$

Progress check

1. The variable X follows a binomial distribution where $X \sim B(n, p)$.

   (a) If $n = 12$ and $p = 0.25$, use cumulative binomial tables to determine:

        (i) $P(X < 5)$,   (ii) $P(X \geqslant 8)$ (iii) $P(2 \leqslant X \leqslant 7)$

   (b) If $n = 16$ and the mean is 12.8, determine, by using appropriate formulae

        (i) p (ii) $P(X = 11)$

2. Given that $X \sim B(20, 0.3)$ write down:

   (a) the mean of X

   (b) the variance of X

   (c) the standard deviation of X.

2 (a) 6 (b) 4.2 (c) 2.05 (3 s.f.)
(b) (i) 0.8 (ii) 0.00739
1 (a) (i) 0.8424 (ii) 0.0028 (iii) 0.8388

4.4 The normal distribution

After studying this section you should be able to:

- use the normal distribution model to calculate probabilities

LEARNING SUMMARY

The normal distribution

The **normal variable** is a continuous random variable. Its probability function is represented by a bell-shaped curve, symmetric about the mean value μ.

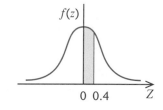

$f(x)$

μ X

The parameters of the normal distribution are μ and σ^2. If the continuous random variable X follows a normal distribution with these **parameters** then this is written as $X \sim N(\mu, \sigma^2)$. The standard deviation is σ.

Tabulated values of probabilities for the normal distribution use Z as the variable.

The **standard normal variable Z** is used to calculate probabilities based on the normal distribution where $Z = \dfrac{X - \mu}{\sigma}$ and $Z \sim N(0, 1)$.

Example

Find P(15 < X < 17) where $X \sim N(15, 25)$

$$15 < X < 17 \implies \frac{15 - 15}{5} < Z < \frac{17 - 15}{5} \implies 0 < Z < 0.4$$

$f(z)$

Use the normal distribution tables (Table 3).

$$P(15 < X < 17) = P(0 < Z < 0.4)$$
$$= \Phi(0.4) - \Phi(0) = 0.65542 - 0.50000$$
$$= 0.15542$$

0   0.4 Z

You may need to use given information to set up and solve simultaneous equations to find the values of μ and σ.

Example

It is given that P(X > 70) = 0.0436 and P(X < 45) = 0.3669, where $X \sim N(\mu, \sigma^2)$. Find the values of μ and σ.

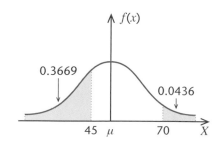

$f(x)$

0.3669

0.0436

45 μ 70 X

$$P(X > 70) = P\left(Z > \frac{70 - \mu}{\sigma}\right) = 0.0436$$

$$\Phi\left(\frac{70 - \mu}{\sigma}\right) = 1 - 0.0436 = 0.9564$$

Approximate 0.9564 to 0.956 (3 d.p.) to look up z in Table 4.

From the tables

$$\left(\frac{70 - \mu}{\sigma}\right) = 1.706 \implies 70 - \mu = 1.706\sigma \qquad [1]$$

Similarly

$$P(X < 45) = P\left(Z < \frac{45 - \mu}{\sigma}\right) = 0.3669$$

$$\Phi\left(\frac{45 - \mu}{\sigma}\right) = 0.3669$$

In general,
$\Phi(-z) = 1 - \Phi(z)$.

$$\Phi\left(-\left(\frac{45 - \mu}{\sigma}\right)\right) = 1 - 0.3669 = 0.6331 = 0.63 \text{ (2 d.p.)}$$

$$-\left(\frac{45 - \mu}{\sigma}\right) = 0.3319$$

$$45 - \mu = -0.3319\sigma \qquad\qquad [2]$$

[1] − [2] gives $25 = 2.0379 \Rightarrow \sigma = 12.267... = 12.3$ (1 d.p.)

Substituting in (1): $\mu = 49.07... = 49.1$ (1 d.p.)

Progress check

1 Find $P(17 < X < 24)$ where $X \sim N(14, 16)$.

2 It is given that $X \sim N(\mu, \sigma^2)$ and $P(X < 30) = 0.7$.
   Given also that $P(X > 35) = 0.1$, find μ and σ.

1 0.2204
2 $\mu = 26.5$
   $\sigma = 6.60$

4.5 Correlation and regression

After studying this section you should be able to:

- *calculate and interpret the product-moment correlation coefficient*
- *recognise explanatory and response variables*
- *find the equations of least squares regression lines*
- *use a suitable regression line to estimate a value*
- *interpret a and b in the equation of the regression line of y on x*

LEARNING SUMMARY

All of the work in this section relates to the treatment of **bivariate data**. This is data in which each data point is defined by two variables.

Correlation

A **scatter diagram** may be used to represent bivariate data. The extent to which the points approximate to a straight line gives an indication of the strength of a linear relationship between the variables, known as the **linear correlation**.

> You should only use this method when both variables are normally distributed.

One way to arrive at a numerical measure of the correlation is to use the **product–moment correlation coefficient**, r.

For n pairs of (x, y) values:

$$S_{xx} = \Sigma(x - \bar{x})^2 = \Sigma x^2 - \frac{(\Sigma x)^2}{n} \qquad S_{yy} = \Sigma(y - \bar{y})^2 = \Sigma y^2 - \frac{(\Sigma y)^2}{n}$$

$$S_{xy} = \Sigma(x - \bar{x})(y - \bar{y}) = \Sigma xy - \frac{(\Sigma x)(\Sigma y)}{n}$$

> When you have actual data, you can use your calculator in REG mode to find r.

and the product–moment correlation coefficient is given by:

$$r = \frac{S_{xy}}{\sqrt{S_{xx}S_{yy}}}$$

Regression

This gives values of r between -1 (representing a perfect negative correlation) and $+1$ (representing a perfect positive correlation).

Whereas correlation is determined by the *strength* of a linear relationship between the two variables, **regression** is about the *form* of the relationship given by the equation of a **regression line**. (Make sure you know the difference between correlation and regression.)

> The response variable is the **dependent** variable. The explanatory variable is the **independent** variable.

The purpose in establishing the equation of a regression line is to make predictions about the values of one variable (known as the **response variable**) for some given values of the other variable (known as the **explanatory variable**).

Predictions should only be made for values within the range of readings of the explanatory variable. **Extrapolation** for values outside this range is unreliable. Another factor affecting the accuracy of any predictions is the influence of **outliers** on the equation of the regression line.

Figure 1 illustrates **y-residuals** given by

$$d = \text{(observed value of } y) - \text{(predicted value of } y).$$

Figure 2 illustrates **x-residuals** given by

$$d = \text{(observed value of } x) - \text{(predicted value of } x).$$

A **least squares regression line** is a line for which the sum of the squares of either the x-residuals or y-residuals is minimised.

> Both lines always pass through the point $(\bar{x}, \bar{y})$.

Figure 1

$\sum d^2 = \min$

This gives the regression line of y on x as

$$y = a + bx,$$

where $b = \dfrac{S_{xy}}{S_{xx}}$ and $a = \bar{y} - b\bar{x}$.

> When you know the actual data, you can find a and b directly from your calculator in REG mode.

Use this equation to estimate values of y for given values of x when x is the explanatory variable and y is the response variable.

Figure 2

$\sum d^2 = \min$

This gives the regression line of x on y as

$$x = c + dy,$$

where $d = \dfrac{S_{xy}}{S_{yy}}$ and $c = \bar{x} - d\bar{y}$.

Use this equation to estimate values of x for given values of y when y is the explanatory variable and x is the response variable.

Unless there is perfect correlation between the variables, the two regression lines will be different and you cannot rearrange one equation to obtain the other.

In the regression line of y on x, which has equation $y = a + bx$:
a is the predicted y value when $x = 0$, i.e. where the line crosses the y-axis.
b is the gradient of the regression line. It gives the average rate at which y is changing with x.

Progress check

1 In each part below, find:
   (a) the value of the product–moment correlation coefficient between x and y.
   (b) the equation of the least squares regression line of y on x.
   (i)

x	12	15	17	24	11	26	32	14
y	102	97	82	70	93	84	70	100

(ii) The summary data for 10 pairs of (x, y) values is as follows:

$$\sum x = 146, \quad \sum x^2 = 2208, \quad \sum y = 147, \quad \sum y^2 = 2247, \quad \sum xy = 2211.$$

4.6 Estimation and sampling

After studying this section you should be able to:

- *find unbiased estimates for population parameters from a sample*
- *understand the theory relating to sampling distributions*
- *use the central limit theorem*
- *find confidence intervals*

LEARNING SUMMARY

Unbiased estimates

When a **population parameter**, such as the mean or variance, is **unknown**, then it is sensible to estimate it from a sample.

An **unbiased estimate** is one which, on average from many samples, gives the true value.

If a distribution has mean μ and variance σ^2, then:

An unbiased estimate of μ is: $\overline{X} = \dfrac{\Sigma x}{n}$

An unbiased estimate of σ^2 is: $S^2 = \dfrac{\Sigma(x_i - \overline{X})^2}{n-1}$

Sampling distribution

The most important sampling distribution is the distribution of the sample mean and you will need to be able to use it to find probabilities.

The sampling distribution of the mean

If all possible samples of size n are taken from $X \sim N(\mu, \sigma^2)$ and their sample means calculated, then these means form the sampling distribution of means, $\overline{X}$ which is also normally distributed.

> This important result can be derived using expectation algebra (See Revise A2).

> **KEY POINT**
>
> If $X \sim N(\mu, \sigma^2)$, then $\overline{X} \sim N\left(\mu, \dfrac{\sigma^2}{n}\right)$.

> The mean of $\overline{X}$ is the same as the mean of X.

> The variance of $\overline{X}$ is much smaller than the variance of X.

> Standard deviation = $\sqrt{\text{variance}}$

The diagram shows the curves for X and for $\overline{X}$.

They are both symmetrical about μ.
The curve for $\overline{X}$ is much more squashed in, confirming the smaller standard deviation.

The standard deviation of the sampling distribution, $\sigma/\sqrt{n}$, is known as the **standard error of the mean**.

The following result is extremely useful and should be learnt.

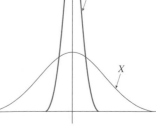

> **KEY POINT**
>
> If the distribution of X is not normal, then, by the **central limit theorem**, $\bar{X} \sim N\left(\mu, \dfrac{\sigma^2}{n}\right)$ approximately, provided n is large.

> Sometimes $\hat{\sigma}$ is written for s.

> s^2 is the unbiased estimate of σ^2.

If you do not know σ^2, the variance of X, then it can be estimated by s^2 and provided n is large, $\bar{X} \sim N\left(\mu, \dfrac{s^2}{n}\right)$ approximately.

Example

The heights of men in a particular area are normally distributed with mean 176 cm and standard deviation 7 cm. A random sample of 50 men is taken. Find the probability that the mean height of the men in the sample is less than 175 cm.

If X is the height, in centimetres, of a man from this area, then $X \sim N(176, 7^2)$.

For random samples of size 50, $\bar{X} \sim N\left(176, \dfrac{7^2}{50}\right)$, i.e. $\bar{X} \sim N(176, 0.98)$.

> In this example, since the population is normal, any size sample, large or small, could have been taken.

$$P(\bar{X} < 175) = P\left(Z < \dfrac{175 - 176}{\sqrt{0.98}}\right)$$
$$= P(Z < -1.01)$$
$$= 1 - \Phi(1.01)$$
$$= 1 - 0.84375$$
$$= 0.156 \ (3 \text{ d.p.})$$

$\bar{X}:$ 175   176
$Z:$ —1.01   0

Confidence intervals

A **confidence interval** for an unknown population parameter is an interval (a, b) such that there is a specified probability, often 90%, 95% or 99%, that the interval contains the true value.

Confidence interval for the mean μ

This is calculated using the mean $\bar{x}$, of a random sample size n.
If the population is **normal**, the sample size n can be **any size**. If it is **not normal**, the sample size n must be large ($n > 30$ say).

> ±1.96 are the standardised z-values for the central 95% of a standard normal distribution.

> Provided n is large, use s if σ is unknown.

> The 95% **confidence limits** for μ are $\bar{x} \pm 1.96 \dfrac{\sigma}{\sqrt{n}}$.
>
> The 95% **confidence interval** is $\left(\bar{x} - 1.96 \dfrac{\sigma}{\sqrt{n}}, \ \bar{x} + 1.96 \dfrac{\sigma}{\sqrt{n}}\right)$

$\bar{x} - 1.96 \dfrac{\sigma}{\sqrt{n}}$ $\bar{x}$ $\bar{x} + 1.96 \dfrac{\sigma}{\sqrt{n}}$

$\longleftarrow$ $2 \times 1.96 \dfrac{\sigma}{\sqrt{n}}$ $\longrightarrow$

The **width** of the 95% confidence interval is $2 \times 1.96 \dfrac{\sigma}{\sqrt{n}}$.

Different levels of confidence

Any level of confidence can be used. The most common are 95% (described above), 90% and 99%.

90% confidence interval is $\left(\bar{x} - 1.6449\,\dfrac{\sigma}{\sqrt{n}}, \bar{x} + 1.6449\,\dfrac{\sigma}{\sqrt{n}}\right)$.

99% confidence interval is $\left(\bar{x} - 2.5758\,\dfrac{\sigma}{\sqrt{n}}, \bar{x} + 2.5758\,\dfrac{\sigma}{\sqrt{n}}\right)$.

Example

X is the mass, in grams, of a bag of flour packed by a particular machine. The mean mass of bags packed by this machine is μ and the standard deviation is σ.
A random sample of 100 bags gave the following results:

$$\Sigma x = 50\,140, \quad \Sigma (x - \bar{x})^2 = 26\,383.5$$

Find a 95% confidence interval for μ.

$$\bar{x} = \frac{\Sigma x}{n} = \frac{50\,140}{100} = 501.4$$

$$s = \sqrt{\frac{\Sigma(x - \bar{x})^2}{n - 1}} = \sqrt{\frac{26383.5}{99}} = 16.32\ldots$$

95% confidence limits are $\bar{x} \pm 1.96\,\dfrac{s}{\sqrt{n}} = 501.4 \pm 1.96\,\dfrac{16.32}{\sqrt{100}} = 501.4 \pm 3.20$

95% confidence interval $= (501.4 - 3.20, 501.4 + 3.20)$
$= (498.2\text{ g}, 504.6\text{ g})$ (1 d.p.)

The distribution of X is not known. Since n is large, the central limit theorem has been used in the underlying theory.

Progress check

1 The random variable X has mean μ and standard deviation σ.

Ten independent observations of X are
4.6, 3.9, 4.2, 5.8, 6.3, 4.2, 7.2, 6.0, 7.1, 5.4.

(a) Find unbiased estimates of μ and σ.
(b) Given that X is normally distributed with variance 2, calculate a 99% confidence interval for μ.

2 A random sample of size 100 is taken from a normal distribution with mean 120 and standard deviation 5.
(a) Find P$(118.75 < \bar{X} < 120.5)$, where $\bar{X}$ is the sample mean.
(b) State, with a reason, whether your answer would be different if X is not normally distributed.

2 (a) 0.8351; (b) no, use central limit theorem since n is large.
1 (a) 5.47, 1.21 (b) (4.32, 6.62)

Sample questions and model answers

1

Paul swims lengths of a swimming pool in training for a competition. His times, x seconds, for completing a sample of 20 lengths of the pool are summarised below:

$$\Sigma x = 372 \quad \Sigma(x - \bar{x})^2 = 31.1296$$

(a) Find the mean and standard deviation of Paul's times for completing one length of the pool.

(b) A month ago, in a random sample of lengths, Paul's mean time per length was 19.8 s with a standard deviation of 1.72 s. Comment on the change in his performance indicated by these results.

Scientific calculators can find the standard deviation from the raw data, but you need to know the formulae because the data may be given in summarised form.

(a)
$$\bar{x} = \frac{\Sigma x}{n} = \frac{372}{20}$$

$$= 18.6.$$

Paul's mean time per length is 18.6 s.

Quote the formula and substitute the information.

$$s = \sqrt{\frac{\Sigma(x - \bar{x})^2}{n - 1}} = \sqrt{\frac{31.1296}{19}}$$

$$= 1.28$$

The standard deviation of Paul's times is 1.28 s.

(b) The mean time has been reduced, suggesting an improvement in performance. The standard deviation has also been reduced which suggests a greater consistency of performance.

You may be expected to interpret results within the context of the question and make comparisons.

2

A machine produces nails with lengths that are normally distributed with mean 5.2 cm and standard deviation 0.06 cm.

Find the probability that a nail selected at random has a length between 5.1 cm and 5.25 cm.

Define the random variable X.

Let X be the length, in cm, of a nail, where $X \sim N(5.2, 0.06^2)$.

Express the probability in terms of X and then standardise.

$$P(5.1 < X < 5.25) = P\left(\frac{5.1 - 5.2}{0.06} < Z < \frac{5.25 - 5.2}{0.06}\right)$$

$$= P(-1.67 < Z < 0.83)$$

$$= \Phi(0.83) - \Phi(-1.67)$$

$$= \Phi(0.83) - (1 - \Phi(1.67))$$

It follows from the symmetry of the distribution that

$$\Phi(-z) = 1 - \Phi(z).$$

$$= 0.79673 - (1 - 0.95254)$$

$$= 0.74927$$

The probability that a nail selected at random has a length between 5.1 cm and 5.25 cm is 0.749 (3 s.f.)

Sample questions and model answers (continued)

3

In an advanced level examination taken by a large number of candidates, the marks were distributed normally with mean mark 68.7 and standard deviation 5.4.

A random sample of 100 scripts is taken and their mean mark denoted by $\bar{X}$.

(a) State the distribution of $\bar{X}$.

(b) Find the probability that the mean mark of the 100 scripts is between 68 and 70, giving your answer correct to 2 significant figures.

Let X be the examination mark, so $X \sim N(68.7, 5.4^2)$

(a) For samples of size 100,

$$\bar{X} \sim N\left(68.7, \frac{5.4^2}{100}\right), \text{ i.e. } \bar{X} \sim N(68.7, 0.2916)$$

(b) $P(68 < \bar{X} < 70) = P\left(\dfrac{68 - 68.7}{\sqrt{0.2916}} < Z < \dfrac{70 - 68.7}{\sqrt{0.2916}}\right)$

$= P(-1.30 < Z < 2.41)$

$= \Phi(2.41) - (1 - \Phi(1.30))$

$= 0.99202 - (1 - 0.90320)$

$= 0.89522$

$= 0.895 \ (3 \ s.f.)$

If the degree of approximation is not specified in the question or your examination rubric, then approximate sensibly.

Practice examination questions

1   X is a random variable such that $X \sim \text{B}(20, 0.4)$.

   (a) Find the mean and variance of X.

   (b) Find the probability that the value of X lies between 2 and 4 inclusive.

   (c) Find the probability that X is greater than or equal to 5.

2   The events A and B are such that

$$P(A) = 0.3 \qquad P(B) = 0.6 \qquad P(A \cup B) = 0.72$$

   (a) Show that the events A and B are not mutually exclusive.

   (b) Show that the events A and B are independent.

   (c) Write down $P(A \cup B)'$.

   (d) Find $P(A' \cup B')$.

3   (a) Three cards are drawn at random, without replacement, from a pack of playing cards.
        Calculate the probability that:

   (i)   All three cards are red.
   (ii)   All three cards are the same colour.
   (iii) At least one of the cards is red.

   (b) The cards are returned to the pack and the pack is shuffled. One more card is drawn.
        Find the probability that this card is an ace, given that it is not a diamond.

4   The lifetime of batteries used in a torch can be modelled by a normal distribution with a mean of 8 h and a standard deviation of 25 mins.

   (a) Calculate the probability that a battery selected at random will last for more than $8\frac{1}{2}$ h.

   (b) Find the percentage of batteries that will last for less than 7 h.

5   A machine distributes 'lucky pendants' to one in four packets of cereal at random. A box contains 20 packets of the cereal.

   (a) Using tables of cumulative binomial probabilities, or otherwise, find the probability that in a randomly selected box:

   (i)   At least six packets contain a lucky pendant.
   (ii) Exactly six packets contain a lucky pendant.

   (b) Five boxes are randomly selected. Find the probability that exactly two of the boxes have at least six packets that contain a lucky pendant.

6 Eleven students in the sixth form of a school study both mathematics and physics at A Level. It is thought that their physics scores depend on their maths scores. The scores in an end of term test are shown for 10 of these students.

| Maths (x) | 48 | 39 | 52 | 55 | 71 | 62 | 76 | 54 | 61 | 58 |
| Physics (y) | 41 | 35 | 50 | 47 | 63 | 48 | 70 | 45 | 46 | 39 |

Using x to represent the score in maths and y to represent the corresponding score in physics:

(a) Calculate the product–moment correlation coefficient, between x and y.

(b) Find the equation of the regression line of y on x for the data.

(c) The eleventh student obtained 65 on the maths test. Estimate this student's score in the physics test, commenting on the reliability of your estimate.

Chapter 5

Decision Mathematics 1

The following topics are covered in this chapter:

- Algorithms
- Graphs and networks
- Linear programming
- Matchings

5.1 Algorithms

After studying this section you should be able to:

- implement an algorithm given by text or flow chart
- implement the following algorithms from memory:
  - bubble sort
  - quick sort
  - shuttle sort
  - shell sort

Introducing algorithms

Algorithms play a fundamental role throughout Decision Mathematics.

In the exam you may be asked to implement an unfamiliar algorithm and interpret the results.

An **algorithm** is a finite sequence of precise instructions used to solve a problem.

One way to define an algorithm is to simply list all of the necessary steps. In more complex situations, the process may be easier to follow if the algorithm is presented as a **flow diagram**.

You need to be able to implement an algorithm presented in either form.

Example

Implement the algorithm given by the flow chart and state what the written values represent.

The table shows how the values of I and J change as the algorithm is implemented.

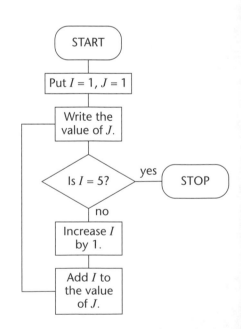

I	1	2	3	4	5
J	1	3	6	10	15

The written values are 1, 3, 6, 10, 15 and these are the first five triangle numbers.

The algorithm may be expressed in words as:

Write down a sequence of 5 numbers starting with 1 and then increasing, first by 2 and then by 1 more each time from term to term.

This version is more concise but not so easy to understand.

In the exam you may be expected to implement any of these algorithms without being reminded of the necessary steps.

You need to know and be able to implement the following algorithms.

The bubble sort

As its name suggests, the **bubble sort** is an algorithm for sorting a list in a particular order.

Step 1   Compare the first two items of the list and switch them if they are in the wrong order.

Step 2   Compare the second and third items of the list and, again, switch them if they are in the wrong order.

Step 3   Continue in the same way until you reach the end of the list. This completes the first **pass** through the list.

Step 4   Make repeated passes through the list until a pass produces no change.

Example   Sort the list 5, 7, 3, 11, 6, 8 into ascending order.

The first pass gives

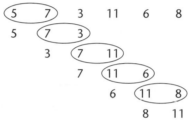

Don't just look at the list and say that the numbers are now in order. Use the algorithm to decide when to stop.

The second pass produces   3, 5, 6, 7, 8, 11.

The third pass produces   3, 5, 6, 7, 8, 11.

The third pass made no change so the ordering is complete.

The shuttle sort

In the shuttle sort, a pass involves comparing one new pair of items and, if a swap is made, shuttling back to compare previous items as detailed below.

Step 1   In the first pass, compare the first two items in the list and swap if necessary.

Step 2   In the second pass, compare the second and third items and swap if necessary. If a swap is made, compare the first and second items and swap if necessary.

Step 3   In the third pass, compare the third and fourth items and swap if necessary. If a swap is made, shuttle back and compare the second and third items, swapping if necessary. If a swap is made shuttle back again and compare the first and second items, swapping if necessary.

Step 4   For a list of n items, continue in the same way until the $(n-1)$th pass is complete.

Example Sort the list 5, 7, 3, 11, 6, 8 into ascending order.
Find the number of comparisons and the number of swaps in the fourth pass.

This shows the list at the **end** of the first pass. The underlined figures are the ones involved in the next pass.

	5	7	3	11	6	8
First pass	5	7	3	11	6	8
Second pass	3	5	7	11	6	8
Third pass	3	5	7	11	6	8
Fourth pass	3	5	6	7	11	8
Fifth pass	3	5	6	7	8	11

The ordered list is 3, 5, 6, 7, 8, 11.

In the fourth pass: Compare 11 and 6   Swap
                     Compare 7 and 6   Swap
                     Compare 5 and 6   Don't swap
In the fourth pass, there are 3 comparisons and 2 swaps.

The shell sort

Step 1   For a list of n items, work out $N_1 = n \div 2$, ignoring any remainder.

Step 2   Extract a sublist composed of the first item and every subsequent N_1th item from the original list.

Step 3   Sort the sublist using the shuttle sort and insert the sorted items back into the spaces in the list.

Step 4   Extract a new sublist starting from the second item and every subsequent N_1th item from the original list.

Step 5   Continue in the same way until every item in the original list has been considered. This completes the first pass.

Step 6   Work out $N_2 = N_1 \div 2$ ignoring any remainder.

Step 7   Extract a sublist composed of the first item and every subsequent N_2th item from the original list.

Step 8   Continue as for N_1 to complete the second pass.

Step 9   Continue with N_3, N_4 as necessary until the final pass is completed with $N_k = 1$.

The shell sort works most efficiently when n is a power of 2.

Example Use the shell sort algorithm to put the following names into alphabetical order: Margaret Dyer, Marius Kowalski, Gemma Leedham, Annette Sherratt, John Wyatt, Paul Arrowsmith, Keith Evans, Simran Hasaan.

Using the first letter of each surname, the original list is:

D	K	L	S	W	A	E	H
D				W			
	A					K	
		E				L	
			H				S
D	A	E	H	W	K	L	S
D		E		L		W	
	A		H		K		S
D	A	E	H	L	K	W	S
A	D	E	H	K	L	S	W

$N_1 = 8 \div 2 = 4$
Positions 1 and (1 + 4) give the first sublist.

$N_2 = 4 \div 2 = 2$

In alphabetical order, the names are: Paul Arrowsmith, Margaret Dyer, Keith Evans, Simran Hasaan, Marius Kowalski, Gemma Leedham, Annette Sherratt, John Wyatt.

The quick sort

Step 1 Take the first item as the **pivot**.

Step 2 Consider each of the other items in turn and place any with order less than or equal to the pivot on its left and any greater than the pivot on its right. Do this without re-ordering the numbers on either side of the pivot.

This creates two sub-lists with the pivot in between.

Step 3 Repeat steps 1 and 2 for each sub-list until each one contains at most one item.

Example Sort the list 4, 2, 6, 10, 5, 7 into ascending order.

$$\underline{4} \quad 2 \quad 6 \quad 5 \quad 7 \quad 10$$

$$2 \quad \boxed{4} \quad \underline{6} \quad 5 \quad 7 \quad 10$$

$$2 \quad \boxed{4} \quad 5 \quad \boxed{6} \quad \underline{7} \quad 10$$

$$2 \quad \boxed{4} \quad 5 \quad \boxed{6} \quad \boxed{7} \quad 10$$

Highlighted values have been used as pivots.

It is easy to see that the list is now in order, but the sub-list 7, 10 has more than one item, so continue with the algorithm.

The ordered list is 2, 4, 5, 6, 7, 10.

Progress check

1 Write the letters P, Q, B, C, A, D in alphabetical order using:

(a) the bubble sort algorithm

(b) the quick sort algorithm.

2 Use the shell sort algorithm to put the numbers 5, 7, 8, 11, 19, 16, 9, 6 into ascending order.

2 5, 6, 7, 8, 9, 11, 16, 19
1 A, B, C, D, P, Q.

5.2 Graphs and networks

After studying this section you should be able to:

- *understand the use of terminology associated with graphs and networks*
- *use Prim's and Kruskal's algorithms for constructing a minimum spanning tree*
- *use Dijkstra's algorithm to find a shortest path between two vertices*
- *find upper and lower bounds for the travelling salesperson problem*
- *use the route inspection algorithm to find a route of minimum weight that traverses every arc of a network*

LEARNING SUMMARY

Defining terms

> It is worth spending time getting to know all of the terms. Their definitions are quite detailed so you will need to keep reminding yourself of what each term means.

- A **graph** is a set of points, called **vertices** or **nodes**, connected by lines called **edges** or **arcs**.
- A **simple graph** is one that has no loops and in which no pair of vertices are connected by more than one edge.

A simple graph

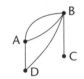
A non-simple graph
Two edges between A and B

A non-simple graph
A loop around C

- The number of edges incident on a vertex is called its **order**, **degree** or **valency**. A vertex may be **odd** or **even** depending on whether its order is odd or even.
- A **subgraph** of some graph G is a graph consisting entirely of vertices and edges that belong to G.
- A **directed** edge is an edge that has an associated direction shown by an arrow. A **digraph** is a graph in which the edges are directed.
- A **path** is a finite sequence of edges such that the end vertex of one edge is the start vertex of the next edge. No edge is included more than once.
- A **connected graph** is one in which every pair of vertices is connected by a path.
- A **complete graph** is one in which every pair of vertices is connected by an edge. A complete graph with n vertices is denoted by K_n.
- A **planar graph** is one that can be drawn in a plane such that no two edges meet except at a vertex. K_4 is planar but K_5 is not planar.

K_4 K_5

- A **cycle** is a path that starts and finishes at the same vertex.
- An **Eulerian** cycle is one that traverses all of the edges of the graph.
- A **Hamiltonian** cycle is one that passes through every vertex once only.
- A **tree** is a graph with no cycles.
- A **spanning tree** is a tree whose vertices are all of the vertices of the graph.
- A **network** is a graph that has a number (**weight**) associated with every edge.
- A **minimum spanning tree** (MST) of a network is a spanning tree of minimum possible weight. It is sometimes called a **minimum connector**.

Prim's algorithm

Prim's algorithm may be used to find a minimum spanning tree of a network.

Step 1   Choose a starting vertex.

Step 2   Connect it to the vertex that will make an edge of minimum weight.

Step 3   Connect one of the remaining vertices to the tree formed so far, in such a way that the minimum extra weight is added to the tree.

Step 4   Repeat step 3 until all of the vertices are connected.

The network given above may be represented by the matrix shown.

Starting from A gives the MST as

AD could have been used instead of CD

> The number in row A and column B, for example, represents the weight of the edge AB.

A version of Prim's algorithm may be used to find the minimum spanning tree directly from a matrix.

Step 1   Choose a starting vertex. Delete the corresponding row and write 1 above the corresponding column as a label.

Step 2   Circle the smallest undeleted value in the labelled column and delete the row in which it lies.

Step 3   Label the column, corresponding to the vertex of the deleted row, with the next label number.

> If there is more than one smallest value then you can choose which one to circle.

Step 4   Circle the smallest undeleted value of all the values in the labelled columns and delete the row in which it lies.

Step 5   Repeat steps 3 and 4 until all of the rows are deleted. The circled values then define the edges of the minimum spanning tree.

	A	B	C	D
A	–	5	–	–
B	5	–	7	–
C	–	7	–	14
D	–	–	14	–

	1 A	2 B	3 C	D
A	–	5	–	–
B	(5)	–	7	–
C	–	(7)	–	14
D	–	–	(14)	–

The circled values correspond to the edges AB, BC, and CD.

Kruskal's algorithm

Kruskal's algorithm is an alternative way to find a minimum spanning tree. In this case, the subgraph produced may not be connected until the final stage.

Step 1   Choose an edge with minimum weight as the first subgraph.

Step 2   Find the next edge of minimum weight that will not complete a cycle when taken with the existing subgraph. Include this edge as part of a new subgraph.

Step 3   Repeat step 2 until the subgraph makes a spanning tree.

BD cannot be included

Dijkstra's algorithm

Dijkstra's algorithm is used to find the shortest distance between a chosen *start* vertex and any other vertex in a network.

The algorithm is designed for use by computers but, when implementing it without the aid of a computer, a system of labelling is required.

Order of labelling

Label (the minimum distance from the start vertex).

Record your 'Working values' here

The steps in the algorithm below refer to this system of labelling. Every vertex is labelled in the same way.

Step 1   Give the start vertex a label of 0 and write its order of labelling as 1.

Step 2   For each vertex directly connected to the start vertex, enter the distance from the start vertex as a working value. Enter the smallest working value as a label for that vertex. Record the order in which it was labelled as 2.

Step 3   For each vertex directly connected to the one that was last given a label, add the edge distance onto the label value to obtain a total distance. This distance becomes the working value for the vertex unless a lower one has already been found.

Step 4   Find the vertex with the smallest working value not yet labelled and label it. Record the order in which it was labelled.

Step 5   Repeat steps 3 and 4 until the target vertex is labelled. The value of the label is the minimum distance from the start vertex.

Step 6   To find the path that gives the shortest distance, start at the target vertex and work back towards the start vertex in such a way that an edge is only included if its distance equals the change in the label values.

Example
Use Dijkstra's algorithm to find the shortest distance from A to F in this network and state the route used.

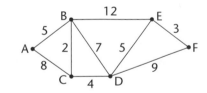

Using Dijkstra's algorithm produces the diagram on the right.

The shortest distance from A to F is 19 (given by the label at F).

Tracing the route backwards using Step 6 gives F, E, D, C, B, A, so the required route is A, B, C, D, E, F.

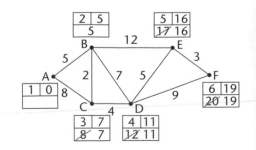

The travelling salesperson problem

This is based on the classic situation of a salesperson who wishes to visit a number of towns and return home using the shortest possible route.

The **travelling salesperson problem** (TSP) is the problem of finding a route of minimum distance that visits every vertex and returns to the start vertex. For a small network it is possible to produce an exhaustive list of all possible routes and choose the one that minimises the total distance. For a large network an exhaustive check is not feasible, even with a computer, because the number of possible routes grows so rapidly.

It is useful to know within what limits the total distance must lie and there are algorithms that can be used for this purpose.

Finding an upper bound

> An upper bound for the total distance involved in the travelling salesperson problem is given by twice the minimum spanning tree.
>
> **KEY POINT**

The **minimum spanning tree** (MST) for the network ABCD was found earlier.

The total length of the MST is 5 + 7 + 14 = 26.

An upper bound for the TSP is 2 × 26 = 52. This corresponds to the route ABCDCBA.

In some cases you might be able to find more than one short-cut that will enable you to reduce the upper bound.

However, an improved (smaller) upper bound can be found by using a **short-cut** from D directly back to A. The route is then ABCDA and the corresponding upper bound is 40.

A different approach is to use the **nearest neighbour algorithm**.

Step 1   Choose a starting vertex.

Step 2   Move from your present position to the nearest vertex not yet visited.

Step 3   Repeat step 2 until every vertex has been visited. Return to the start vertex.

For the network ABCD considered above, the nearest neighbour algorithm gives the same result as using the MST with a short-cut. This won't always be the case.

Finding a lower bound

Step 1   Choose a vertex and delete it along with all of the edges connected to it.

Step 2   Find a minimum spanning tree for the remaining part of the network.

Step 3   Add the length of the MST to the lengths of the two shortest deleted edges.

The value found at step 3 is a lower bound for the travelling salesperson problem. It may be possible to find a better (larger) lower bound by choosing to delete a different vertex initially and repeating the process.

Deleting D to start with gives a lower bound of 40. This corresponds to a cycle ABCDA and so represents the solution of the TSP.

Using the network ABCD again and deleting A gives the minimum spanning tree BCD of length 7 + 14 = 21. Adding the two shortest deleted lengths gives a lower bound of 37.

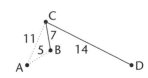

The route inspection problem

> Again, it is common to use length rather than weight in this context.

The **route inspection problem** is to find a route of minimum total length that traverses every edge of the network, at least once, and returns to the start vertex.

This is also known as the Chinese postman problem.

> Leonhard Euler was the most prolific mathematician of all time. One of his many contributions to the subject was to introduce graph theory as a means of solving problems.

The algorithm for solving this problem is based on the idea of a **traversable** graph. A graph is traversable if it can be drawn in one continuous movement without going over the same edge more than once. If all of the vertices of the graph are even, then any one of them may be used as the starting point and the same point will be returned to at the end of the movement. Such a graph is said to be **Eulerian**. The only other possibility for a traversable graph is that it has exactly two odd vertices. In this situation, one of the vertices is the start point and the other is the finish point. This type of graph is said to be **Semi-Eulerian**.

The **route inspection algorithm** is:

Step 1   List all of the odd vertices.

Step 2   Form the list into a set of pairs of odd vertices. Find all such sets.

Step 3   Choose a set. For each pair find a path of minimum length that joins them. Find the total length of these paths for the chosen set.

> Repeating the edges between the pairs of odd vertices in this way effectively makes all the vertices even and creates a traversable graph.

Step 4   Repeat step 3 until all sets have been considered.

Step 5   Choose the set that gives the minimum total. Each pair in the set defines an edge that must be repeated in order to solve the problem.

Progress check

1 (a) Use Prim's algorithm to find the total weight of a minimum spanning tree for this network.

   (b) Verify your answer to part (a) using Kruskal's algorithm.

2 Use Dijkstra's algorithm to find the shortest distance from A to G in the network given in question 1. State the route used.

5.3 Linear programming

After studying this section you should be able to:

- formulate a linear programming problem in terms of decision variables
- use a graphical method to represent the constraints and solve the problem

Formulating a linear programming problem

To formulate a linear programming problem you need to:

- Identify the **variables** in the problem and give each one a label.

x and y are often used for the variables.

- Express the **constraints** of the problem in terms of the variables. You need to include non-negativity constraints such as $x \geqslant 0$, $y \geqslant 0$.

Typically, this may be to maximise a profit or minimise a loss.

- Express the quantity to be optimised in terms of the variables. The expression produced is called the **objective function**.

Example

A small company produces two types of armchair. The cost of labour and materials for the two types is shown in the table.

	Labour	Materials
Standard	£30	£25
Deluxe	£40	£50

The total spent on labour must not be more than £1150 and the total spent on materials must not be more than £1250. The profit on a standard chair is £70 and the profit on a deluxe chair is £100. How many chairs of each type should be made to maximise the profit?

In this case, the variables are the number of chairs of each type that may be produced. Using x to represent the number of standard chairs and y to represent the number of deluxe chairs, the constraints may be written as:

$$30x + 40y \leqslant 1150 \Longrightarrow 3x + 4y \leqslant 115$$

$$25x + 50y \leqslant 1250 \Longrightarrow x + 2y \leqslant 50$$

It's a good idea to simplify the constraints where possible.

and $\qquad x \geqslant 0, \quad y \geqslant 0.$

Using P to stand for the profit, the problem is to maximise $P = 70x + 100y$.

The graphical method of solution

Each constraint is represented by a region on the graph. It's a good idea to shade out the *unwanted* region for each one. The part that remains unshaded then defines the **feasible region** containing the points that satisfy all of the constraints.

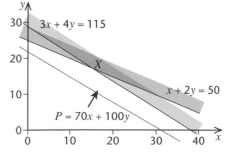

The blue line represents the points where the profit takes a particular value. Moving the line in the direction of the arrow corresponds to increasing the profit. This suggests that the maximum profit occurs at the point X.

X does not represent the solution in this case because both x and y must be integers.

Solving $3x + 4y = 115$ and $x + 2y = 50$ simultaneously gives X as $(15, 17.5)$.

The nearest points with integer coordinates in the feasible region are $(15, 17)$ and $(14, 18)$. The profit, given by $P = 70x + 100y$, is greater at $(14, 18)$.

The maximum profit is made by producing 14 standard and 18 deluxe chairs.

Progress check

1   In the diagram, the shaded region represents the feasible region for a linear programming problem. All of the boundary lines are included.

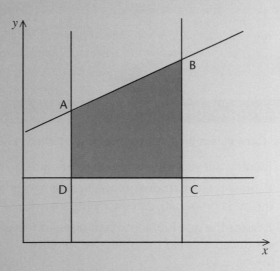

The object is to maximise a profit given by $P = 1.2x + 0.9y$

Which point of the feasible region represents the maximum profit?

1   B (see diagram for explanation)

5.4 Matchings

After studying this section you should be able to:

- *use bipartite graphs to model matchings*
- *understand the conditions for matchings to be maximal or complete*
- *apply the maximum matching algorithm*

LEARNING SUMMARY

Matchings and graphs

A **bipartite graph** is a graph in which the vertices are divided into two sets such that no pair of vertices in the same set is connected by an edge.

In this case, the two sets are {A, B, C} and {p, q, r, s}.

Some vertices in a bipartite graph may not be connected to another vertex.

A **matching** between two sets may be represented by a bipartite graph in which there is at most one edge connecting a pair of vertices.

A **maximal matching** is a matching which has the maximum number of edges. This occurs when every vertex in one of the sets is connected to a vertex in the other set. The bipartite graph shown above represents a maximal matching.

A **complete matching** is a matching in which every vertex is connected to another vertex. This can only occur when the two sets contain the same number of vertices.

The matching improvement algorithm

Figure 1 is a bipartite graph showing the possible connections between two sets. It does not represent a matching because some vertices have more than one connection.

Figure 2 is a bipartite graph representing an initial matching.

An initial matching may be improved by increasing the number of connections. This is the purpose of the **matching improvement algorithm**.

Figure 1

Figure 2

> In an alternating path, the edges alternate between those that are not in the initial matching and those that are.

Step 1 In the initial matching, start from a vertex that is not connected and look for an **alternating path** to a vertex in the other set that is not connected.

Alternating path

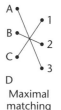

Maximal matching

Step 2 Each edge on the alternating path, not included in the initial matching, is now included and each edge originally included is removed.

Step 3 Repeat steps 1 and 2 using the latest matching in place of the initial matching until no further alternating paths can be found.

Progress check

1 Starting with an initial matching in which 1 is connected to R and 2 is connected to P, use the improvement algorithm to establish a complete matching for this bipartite graph.

1 1–P, 2–S, 3–R, 4–Q.

Sample questions and model answers

1

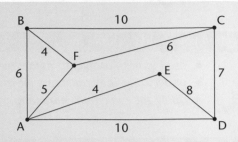

The diagram represents the system of pathways that a security guard must patrol during his course of duty. The weights on the edges represent the time taken, in minutes, to patrol each pathway.

Find the minimum time required to patrol every pathway at least once and give a possible route.

> You need to recognise that this question requires the use of the route inspection algorithm.

> Step 1 of the algorithm is to list the odd vertices.

Vertex	Order
A	4
B	3
C	3
D	3
E	2
F	3

The odd vertices are B, C, D and F.

The sets of pairs of odd vertices are: {BC, DF}
{BD, CF}
{BF, CD}.

> Step 2 of the algorithm is to find the sets of pairs of odd vertices. Each odd vertex appears once in each set.

For the set {BC, DF}

BC = 10. The shortest route from D to F is DC + CF = 13.

> There is an edge connecting B and C.

Total length of the extra paths for this set is 10 + 13 = 23.

> Step 3 of the algorithm is to find the extra length introduced for each set.

For the set {BD, CF}

The shortest route from B to D is BA + AD = 16. CF = 6.

Total length of the extra paths for this set is 16 + 6 = 22.

For the set {BF, CD}

> There is an edge connecting C and F.

BF = 4 and CD = 7.

Total length of the extra paths for this set is 4 + 7 = 11.

The minimum total corresponds to the set {BF, CD} which means that the edges BF and CD will be repeated.

> Step 4 of the algorithm is to keep going until all the sets have been considered.

> There are many possible routes.

A possible route is ABFCDEAFBCDA.

The time taken will be 60 + 11 = 71 minutes.

Practice examination questions

1 A new theme park has its major attractions at A, B, C, D, E, F and G as shown in the network below. The edges of the network show the *possible* routes of paths connecting the attractions.

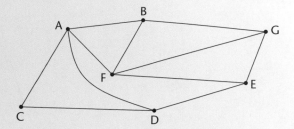

The cost of laying each of these paths, in hundreds of pounds, is given in the table:

	A	B	C	D	E	F	G
A	–	50	40	61	–	38	–
B	50	–	–	–	–	47	54
C	40	–	–	35	–	–	–
D	61	–	35	–	42	–	–
E	–	–	–	42	–	25	10
F	38	47	–	–	25	–	63
G	–	54	–	–	10	63	–

The paths that are actually laid must form a connected network.
Use Prim's algorithm, starting by deleting row A, to find the minimum cost of laying the necessary paths.

2

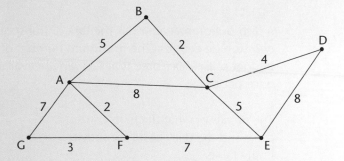

(a) Use Dijkstra's algorithm to find the shortest distance from G to D.

(b) Describe the route taken.

Practice examination questions (continued)

3

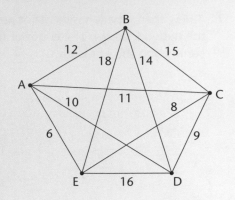

(a) Use Kruskal's algorithm to find the length of the minimum spanning tree for this network.

State the order in which you include each of the edges.

(b) Find an upper bound for the travelling salesperson problem using the minimum spanning tree with shortcuts.

4

(i)

(ii)

(iii)

(a) List the order of the vertices for each graph.

(b) Describe each of the graphs as Eulerian, Semi-Eulerian or neither.

5 A distribution manager has the task of transporting pallets of frozen food between two storage centres. The maximum number of pallets is to be delivered within a daily budget of £2000. Three types of van are available and each van can only do the trip once per day.

The details of what the vans can carry and the daily costs are shown in the table.

At most 10 drivers may be used in one day.

Van type	No. of pallets	Cost/day
Class A	4	£120
Class B	9	£300
Class C	12	£400

Let a represent the number of Class A vans used, b the number of Class B vans used, c the number of Class C vans used and P the total number of pallets transported in a day.

(a) Write an expression for the objective function.

(b) Formulate the task as a linear programming problem.

(c) Describe any special condition that the solution must satisfy.

Practice examination questions *(continued)*

6 The table shows the names of five employees of a company and the days when they are each available to work a late shift. The days that have been highlighted represent a first attempt at matching the employees to the available days to make a roster.

(a) Use a bipartite graph to represent the relationship between employees and days available.

(b) Use a second bipartite graph to represent an initial matching based on the days highlighted.

(c) Implement a matching improvement algorithm to obtain a complete matching. Indicate the alternating path used.

Name	Days available
Abbie	**Mon**, Tue, Fri
Ben	Mon, **Wed**, Thu
Colin	Mon, **Fri**
Dave	Fri, Wed
Emma	**Thu**, Fri

Practice examination answers

1 (a) $x^2 - 4x + 1 = (x-2)^2 - 3$.

(b) $(x-2)^2 - 3 = 0$

$\Rightarrow x - 2 = \pm\sqrt{3}$

$\Rightarrow x = 2 \pm \sqrt{3}$.

(c) $(2, -3)$

(d)

2 (a) So, $f(x) = (x-2)(x^2 + 6x + 9)$

$= (x-2)(x+3)^2$

(b) (c)

3 The coordinates at the points of intersection are given by the simultaneous solution of:

$y = 2x - 2$ [1]

$y = x^2 - x - 6$ [2]

Eliminating y gives

$x^2 - x - 6 = 2x - 2$

$\Rightarrow \quad x^2 - 3x - 4 = 0$

$\Rightarrow \quad (x+1)(x-4) = 0$

$\Rightarrow \quad x = -1$ or $x = 4$.

When $x = -1$, $y = -4$.

When $x = 4$, $y = 6$.

The points of intersection are A$(-1, -4)$ and B$(4, 6)$.

4 $3x^2 - 8x - 7 < 2x^2 - 3x - 11$

$\Rightarrow x^2 - 5x + 4 < 0$

$\Rightarrow (x-1)(x-4) < 0$

$\Rightarrow 1 < x < 4$.

5 (a) Gradient of AB $= 6/5$.

Using $y - y_1 = m(x - x_1)$ gives

$y - 1 = 6/5(x - 4)$

$\Rightarrow \quad 5y - 5 = 6x - 24$

$\Rightarrow \quad 6x - 5y - 19 = 0$.

(b) The midpoint of A and B is $(1.5, -2)$.

The gradient of the line l is $-5/6$.

Using $y - y_1 = m(x - x_1)$ again gives

$y + 2 = -5/6(x - 1.5)$

$\Rightarrow \quad 6y + 12 = -5x + 7.5$

$\Rightarrow \quad 12y + 24 = -10x + 15$

$\Rightarrow \quad 10x + 12y + 9 = 0$.

(c) At C, $x = 0 \Rightarrow y = -0.75$ so the coordinates of C are $(0, -0.75)$.

6 (a) The equation of the line m is $3x + 2y = 6$

$\Rightarrow \qquad y = -3/2x + 3$

$\Rightarrow \qquad$ The gradient of the line m is $-3/2$

$\Rightarrow \qquad$ The gradient of the line l is $2/3$.

Using $y - y_1 = m(x - x_1)$ gives

$y - 4 = 2/3(x - 5)$

$\Rightarrow \qquad 3y - 12 = 2x - 10$

$\Rightarrow \qquad 2x - 3y + 2 = 0$ which is the equation of line l.

(b) At A, $y = 0 \Rightarrow x = -1$.

B lies on the line with equation $3x + 2y = 6$.

At B, $y = 0 \Rightarrow x = 2$.

So the distance AB is given by $2 - (-1) = 3$.

Core 1 *(continued)*

7 (a) $y = 2x^3 - 15x^2 - 36x + 10$

$$\frac{dy}{dx} = 6x^2 - 30x - 36$$

At a stationary point

$$\frac{dy}{dx} = 0 \Rightarrow 6x^2 - 30x - 36 = 0$$

$$\Rightarrow x^2 - 5x - 6 = 0$$

$$\Rightarrow (x + 1)(x - 6) = 0$$

$$\Rightarrow x = -1 \text{ or } x = 6.$$

When $x = -1$, $y = 29$.

When $x = 6$, $y = -314$.

The stationary points are $(-1, 29)$ and $(6, -314)$.

(b) $\dfrac{d^2y}{dx^2} = 12x - 30$

When $x = -1$, $\dfrac{d^2y}{dx^2} = -42 < 0$ maximum.

When $x = 6$, $\dfrac{d^2y}{dx^2} = 42 > 0$ minimum.

(c) The function is decreasing at all of the points between the turning points, i.e. when $-1 < x < 6$.

8 (a) $x + y = 7$

(b) At P and Q $x + y = 7$ [1]

and $y = (x - 1)^2 + 4$ [2]

Substituting for y in [2] gives

$$7 - x = x^2 - 2x + 5$$

$$\Rightarrow x^2 - x - 2 = 0$$

$$\Rightarrow (x - 2)(x + 1) = 0$$

$$\Rightarrow x = 2 \text{ or } x = -1.$$

$x = 2 \Rightarrow y = 5$ and $x = -1 \Rightarrow y = 8$.

P is the point $(-1, 8)$ and Q is the point $(2, 5)$.

(c) The shaded area is given by

$$\frac{3}{2}(8 + 5) - \int_{-1}^{2} (x^2 - 2x + 5)\, dx$$

$$= 19.5 - 15$$

$$= 4.5$$

9 (a) By the factor theorem $f(-2) = 0$

$$\Rightarrow (-2)^3 - 4(-2)^2 - 3(-2) + k = 0$$

$$\Rightarrow -8 - 16 + 6 + k = 0$$

$$\Rightarrow k = 18.$$

(b) $f(x) = x^3 - 4x^2 - 3x + 18$

$$\equiv (x + 2)(Ax^2 + Bx + C).$$

Comparing coefficients of x^3 gives $A = 1$.

Comparing the constant terms gives $C = 9$.

When $x = -1$

$$-1 - 4 + 3 + 18 = A - B + C$$

$$\Rightarrow 16 = 1 - B + 9 \Rightarrow B = -6.$$

$$f(x) = (x + 2)(x^2 - 6x + 9)$$

$$= (x + 2)(x - 3)^2.$$

(c) From the sketch, $f(x) \geqslant 0 \Rightarrow x \geqslant -2$.

(d) By the remainder theorem, the remainder is given by $f(-1) = 16$.

Core 2

1 (a) $49^{-\frac{1}{2}} = \dfrac{1}{49^{\frac{1}{2}}} = \dfrac{1}{\sqrt{49}} = \dfrac{1}{7}$.

(b) $\dfrac{a \times a\sqrt{a^3}}{\sqrt{a}} = \dfrac{a \times a^{\frac{3}{2}}}{a^{\frac{1}{2}}} = \dfrac{a^{\frac{5}{2}}}{a^{\frac{1}{2}}} = a^2$, so $k = 2$.

2 (a) $8^{x-3} = 4^{x+1} \Rightarrow (2^3)^{x-3} = (2^2)^{x+1}$

$$\Rightarrow 2^{3(x-3)} = 2^{2(x+1)}$$

$$\Rightarrow 3(x - 3) = 2(x + 1)$$

$$\Rightarrow x = 11.$$

(b) $5x^{-\frac{1}{3}} = x^{\frac{1}{6}} \Rightarrow 5x^{-\frac{1}{3}} \times x^{\frac{1}{3}} = x^{\frac{1}{6}} \times x^{\frac{1}{3}}$

$$\Rightarrow 5 = x^{\frac{1}{2}}$$

$$\Rightarrow x = 25.$$

3 10th term $= a + 9d = 74$.   [1]

Sum of first 20 terms $= 10(2a + 19d) = 1510$

$$\Rightarrow \qquad 2a + 19d = 151$$ [2]

$2 \times$ [1] gives $2a + 18d = 148$ [3]

[2] − [3] gives $d = 3$

Subs for d in [1] gives $a + 27 = 74 \Rightarrow a = 47$.

First term $= 47$, common difference $= 3$.

Core 2 *(continued)*

4 (a) First term $a = 3$, common ratio $r = 2$.
The 8th term is $ar^7 = 3 \times 2^7 = 384$.

(b) First term $a = 25$, common difference $d = 1.2$
number of terms $n = 50$.

Using $S_n = \dfrac{n}{2}(2a + (n-1)d)$ gives

Sum $= \frac{50}{2}(2 \times 25 + 49 \times 1.2) = 2720$.

5 (a) $3\log_a x - 2\log_a y + \log_a(x+1)$
$= \log_a x^3 - \log_a y^2 + \log_a(x+1)$
$= \log_a\left(\dfrac{x^3(x+1)}{y^2}\right)$.

(b) $5^x = 100$
$\Rightarrow \log_{10} 5^x = \log_{10} 100$
$x\log_{10} 5 = \log_{10} 100$
$x = \dfrac{\log_{10} 100}{\log_{10} 5} = 2.861$ to 3 d.p.

6 $\cos^{-1}(0.4) = 66.42\ldots°$
$0° \leqslant x \leqslant 360° \Rightarrow 30° \leqslant 2x + 30 \leqslant 750°$
Sketch the graph of $y = \cos X$, $30° \leqslant X \leqslant 750°$,
and the graph of $y = 0.4$

From the sketch, the values of X (i.e $2x + 30$) are:
$66.42\ldots°$, $360 - 66.42\ldots°$, $360 + 66.42\ldots°$,
$720 - 66.42\ldots°$,

This gives
$x = 18.2°$ or $x = 131.8°$ or $x = 198.2°$ or $x = 311.8°$

7 (a) $\cos x + 3\sin x \tan x - 2 = 0$
$\Rightarrow \cos^2 x + 3\sin^2 x - 2\cos x = 0$
$\Rightarrow \cos^2 x + 3(1 - \cos^2 x) - 2\cos x = 0$
$\Rightarrow 3 - 2\cos^2 x - 2\cos x = 0$
$\Rightarrow 2\cos^2 x + 2\cos x - 3 = 0$

(b) $\cos x = \dfrac{-2 \pm \sqrt{4 + 24}}{4}$
$\Rightarrow \cos x = 0.8228\ldots$ or $\cos x = -1.8228\ldots$
(No solutions)
$\Rightarrow x = 34.6°$ or $x = 325.4°$

8 (a) $f'(x) = 2x - x^{-2}$
$f'(2) = 4 - \frac{1}{4} = 3\frac{3}{4}$

(b) $f''(x) = 2 + 2x^{-3}$
$f''(-1) = 2 - 2 = 0$.

9 Area $= \displaystyle\int_1^4 3x^{\frac{1}{2}}\, dx$
$= [2x^{\frac{3}{2}}]_1^4 = 2(8 - 1)$
$= 14$.

10 Area sector = area triangle + area segment
$\Rightarrow \dfrac{1}{2}\theta = \dfrac{1}{2}\sin\theta + \dfrac{\pi}{5}$
$\Rightarrow \theta = \sin\theta + \dfrac{2\pi}{5}$.

11 Area $= \displaystyle\int_2^3 \sqrt{x^2 - 3}\, dx$.
$h = \dfrac{3 - 2}{5} = 0.2$
$A = 0.1\{(1 + 2.4495) + 2(1.3565 + 1.6613 + 1.9591 + 2.2)\}$
$A = 1.78$ to 2 d.p.

12 (a) $2x^{\frac{3}{2}}\left(1 - \dfrac{x}{5}\right)$

(b) $(0, 0)$, $(5, 0)$

(c) $\dfrac{dy}{dx} = 3x^{\frac{1}{2}} - x^{\frac{3}{2}}$

(d) $\dfrac{dy}{dx} = 0 \Rightarrow x^{\frac{1}{2}}(3 - x) = 0$
$\Rightarrow x = 0$ or $x = 3$
Since $x > 0$, $x = 3$
when $x = 3$, $y = 4.157$ (4 s.f.)
$\dfrac{d^2y}{dx^2} = \frac{3}{2}x^{-\frac{1}{2}} - \frac{3}{2}x^{\frac{1}{2}}$
when $x = 3$, $\dfrac{d^2y}{dx^2} = \frac{3}{2}(3^{-\frac{1}{2}} - 3^{\frac{1}{2}}) < 0$

So the turning point at $(3, 4.157)$ is a maximum.

Mechanics 1

1 (a)
$$u = 10$$
$$a = -4$$

Using $v = u + at$ $v = 10 - 4 \times 5 = -10$.

The speed of the particle when $t = 5$ is 10 m s^{-1}.

(b)
$$v = 0 \Rightarrow 10 - 4t = 0$$
$$\Rightarrow t = 2.5.$$

Using $s = ut + \frac{1}{2}at^2$

when $t = 2.5$ $s = 10 \times 2.5 - 2 \times 6.25$

     $s = 12.5$.

when $t = 4$ $s = 10 \times 4 - 2 \times 16$

     $s = 8$.

The total distance travelled between $t = 0$ and $t = 4$ is 12.5 m + 4.5 m = 17 m.

2 (a) The acceleration is positive between $t = 0$ and $t = 25$ so the greatest velocity occurs when $t = 25$.

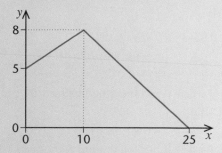

(b) The area under the graph represents the increase in velocity between $t = 0$ and $t = 25$.

Area $= \frac{1}{2} \times 10(5 + 8) + \frac{1}{2} \times 15 \times 8 = 125$.

The increase in velocity is 125 m s^{-1}.

The initial velocity is 20 m s^{-1}.

The maximum value of the velocity is 145 m s^{-1}.

3 (a) The acceleration is constant so the constant acceleration formulae may be used.
$$\mathbf{u} = 2\mathbf{i} + \mathbf{j}$$
$$\mathbf{a} = \mathbf{i} + 3\mathbf{k}$$
$$t = 4.$$

Using $\mathbf{s} = \mathbf{u}t + \frac{1}{2}\mathbf{a}t^2$

     $\mathbf{s} = (2\mathbf{i} + \mathbf{j})4 + (\mathbf{i} + 3\mathbf{k})8$

     $\mathbf{s} = 16\mathbf{i} + 4\mathbf{j} + 24\mathbf{k}$.

This is the displacement from P to Q

so $\overrightarrow{PQ} = 16\mathbf{i} + 4\mathbf{j} + 24\mathbf{k}$.

(b) $\overrightarrow{OQ} = \overrightarrow{OP} + \overrightarrow{PQ}$

     $= -5\mathbf{i} + 11\mathbf{j} + \mathbf{k} + 16\mathbf{i} + 4\mathbf{j} + 24\mathbf{k}$

     $= 11\mathbf{i} + 15\mathbf{j} + 25\mathbf{k}$.

(c) Average speed $= \dfrac{\text{distance}}{\text{time}} = \dfrac{|\overrightarrow{PQ}|}{4}$ m s^{-1}

$|\overrightarrow{PQ}| = \sqrt{16^2 + 4^2 + 24^2} = 29.12$.

Average speed = 7.28 m s^{-1} to 2 d.p.

4 (a)

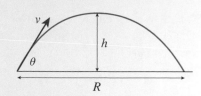

Vertically
$$u = v \sin \theta$$
$$a = -g$$

At max height $v = 0$.

Using $v = u + at$

     $0 = v \sin \theta - gt$

$\Rightarrow$ $t = \dfrac{v \sin \theta}{g}$.

Using $s = ut + \frac{1}{2}at^2$

$h = v \sin \theta . \dfrac{v \sin \theta}{g} - \dfrac{1}{2} g \left(\dfrac{v \sin \theta}{g} \right)^2$

$h = \dfrac{v^2 \sin^2 \theta}{2g}$.

(b) The motion is symmetrical, so total time of flight is 2 × time taken to reach the greatest height.

Time of flight $= \dfrac{2v \sin \theta}{g}$.

Horizontal component of velocity $= v \cos \theta$.

Range $= v \cos \theta \times \dfrac{2v \sin \theta}{g} = \dfrac{v^2(2 \sin \theta \cos \theta)}{g}$

     $= \dfrac{v^2 \sin 2\theta}{g}$.

5

Vertically
$$u = 15 \sin 30°$$
$$a = -9.8$$
$$s = 2.$$

Using $s = ut + \frac{1}{2}at^2$

     $2 = 15 \sin 30° t - 4.9t^2$

$\Rightarrow 4.9t^2 - 15 \sin 30° t + 2 = 0$

$\Rightarrow t = \dfrac{15 \sin 30° \pm \sqrt{(15 \sin 30°)^2 - 8 \times 4.9}}{9.8}$

$\Rightarrow t = 1.186 \ldots$ or $t = 0.3439 \ldots$

(there are two times when the ball has height 2 m)

     $d = 1.186 \ldots \times 15 \cos 30°$

The maximum distance = 15.4 m to 1 d.p.

Mechanics 1 (continued)

6

(a) Resolving vertically

$$R + 50 \sin 40° - 120 = 0$$
$$\Rightarrow R = 120 - 50 \sin 40°$$
$$\Rightarrow R = 87.86 \dots$$

The normal reaction is 87.9 N to 1 d.p.

(b) Resolving horizontally

$$F - 50 \cos 40° = 0$$
$$\Rightarrow F = 38.30 \dots$$

For limiting equilibrium

$$F = \mu R$$

giving

$$38.30 \dots = \mu \times 87.86 \dots$$
$$\mu = 0.4359 \dots$$

The coefficient of friction is 0.436 to 3 d.p.

7

The equations are simpler to deal with by resolving vertically and horizontally, in this case, instead of parallel and normal to the plane.

The minimum value of P corresponds to the case where friction is limiting, as shown in the diagram.

Resolving vertically

$$R \cos 30° + 0.3\,R \sin 30° = 6 \times 9.8$$

$$R = \frac{6 \times 9.8}{\cos 30° + 0.3 \sin 30°}$$

$$R = 57.87 \dots$$

Resolving horizontally

$$P + 0.3R \cos 30° - R \sin 30° = 0$$

$$P = R \sin 30° - 0.3R \cos 30°$$

$$P = 13.90 \dots$$

The minimum value of P is 13.9 to 1 d.p.

8

(a) The equations of motion for the particles are:

For A $\qquad T - 19.6 = 2a$

For B $\qquad 29.4 - T = 3a$.

Adding gives $9.8 = 5a \Rightarrow a = 1.96$

The acceleration of the system is 1.96 m s^{-2}

(b) $T - 19.6 = 3.92 \Rightarrow T = 23.52$.

The tension in the string is 23.52 N.

(c) For B $\quad u = 0$

$\qquad\qquad a = 1.96,$

$\qquad\qquad s = 1.$

Using $\quad s = ut + \dfrac{1}{2}at^2$

$$1 = 0 + 0.98t^2 \ (t > 0)$$
$$\Rightarrow t = 1.01 \dots$$

Particle B hits the ground after 1.01 s to 2 d.p.

(d) Using $\quad v = u + at$

$$v = 0 + 1.96 \times 1.01 \dots = 1.979 \dots$$

Particle B hits the ground with speed 1.98 m s^{-1} to 2 d.p.

(e) Particle A rises to height 2 m then the string becomes slack and it behaves as a projectile.

Using $\qquad\qquad v^2 = u^2 + 2as$

at max height $\qquad 0 = 1.979^2 - 19.6s$

$$\Rightarrow s = 0.2$$

The max height reached by particle A is 2.2 m.

(f) In the extreme situation where the pulley does not move, each particle would be held in equilibrium and so the tension on each side would match the weight of the corresponding particle.

In the given situation, where the pulley is not light or frictionless, it cannot be assumed that the tension on each side of the pulley is the same. This introduces an extra unknown value into the equations. For example, the tension on one side might be labelled T_1 and on the other T_2.

Mechanics 1 *(continued)*

9 Before impact $5u$

After impact v $2v$

Momentum before = momentum after

so $10mu = 2mv + 2mv$

$\Rightarrow 10u = 4v$

$\Rightarrow v = 2.5u$

10 (a) Using $y = x \tan\theta - \dfrac{gx^2}{2v^2}(1 + \tan^2\theta)$

gives $2.5 = 20\tan\theta - \dfrac{9.8 \times 20^2}{2 \times 25^2}(1 + \tan^2\theta)$

$3.136\tan^2\theta - 20\tan\theta + 5.636 = 0$

$\tan\theta = 6.082\ldots$ or $\tan\theta = 0.2954\ldots$

$\tan^{-1}(0.2954\ldots) = 16.5°$

so a possible angle of projection is $16.5°$.

(b) Vertically: using $s = ut + \frac{1}{2}at^2$ gives

$-0.5 = 25\sin 16.5° t - 4.9t^2$

$4.9t^2 - 25\sin 16.5° t - 0.5 = 0$

$t = 1.516\ldots$ or $t = -0.067\ldots$

since $t > 0$

$d = 25\cos 16.5° \times 1.516 - 20$

The distance is 16.3 m to 3 s.f.

Statistics 1

1 $X \sim B(20, 0.4)$ so $n = 20$ and $p = 0.4$

(a) Mean of $X = np = 20 \times 0.4 = 8$

Variance of $X = np(1 - p) = 20 \times 0.4 \times 0.6$

$= 4.8$

(b) $P(2 \leqslant X \leqslant 4) = P(X = 2) + P(X = 3) + P(X = 4)$.

$= \dbinom{20}{2}(0.4)^2(0.6)^{18} + \dbinom{20}{3}(0.4)^3(0.6)^{17}$

$+ \dbinom{20}{4}(0.4)^4(0.6)^{16}$

$= 0.00309 + 0.01235 + 0.03499 = 0.05043$

$= 0.050$ to 3 d.p.

(c) $P(X \geqslant 5) = 1 - P(X \leqslant 4)$.

$= 1 - (P(X = 0) + P(X = 1) + P(2 \leqslant X \leqslant 4))$

$= 1 - (0.6^{20} + 20 \times 0.4 \times 0.6^{19} + 0.05043)$

$= 0.949$ to 3 d.p.

Try checking your answers to (b) and (c) using cumulative binomial tables.

2 (a) $P(A \cup B) = P(A) + P(B) - P(A \cap B)$

$0.72 = 0.3 + 0.6 - P(A \cap B)$

$\Rightarrow P(A \cap B) = 0.18$

As $P(A \cap B) \neq 0$,

A and B are not mutually exclusive.

(b) $P(A) \times P(B) = 0.3 \times 0.6 = 0.18$

As $P(A \cap B) = P(A) \times P(B)$,

A and B are independent.

(c) $P(A \cup B)' = 1 - P(A \cup B) = 0.28$

(d) $P(A' \cup B') = P(A') + P(B') - P(A' \cap B')$

$= 0.7 + 0.4 - 0.7 \times 0.4$

$= 0.82$

3 (a) (i) $P(RRR) = \dfrac{26}{52} \times \dfrac{25}{51} \times \dfrac{24}{50} = \dfrac{2}{17}$

(ii) $P(\text{same colour}) = P(RRR) + P(BBB)$

$= \dfrac{2}{17} + \dfrac{2}{17} = \dfrac{4}{17}$.

(iii) $P(\text{at least one red}) = 1 - P(BBB)$

$= 1 - \dfrac{2}{17} = \dfrac{15}{17}$.

(b) Using A to represent getting an ace and D to represent getting a diamond:

$P(A|D') = \dfrac{P(A \cap D')}{P(D')} = \dfrac{\frac{3}{52}}{\frac{3}{4}} = \dfrac{3}{52} \times \dfrac{4}{3}$

$= \dfrac{1}{13}$.

Statistics 1 *(continued)*

4 Using X to represent the lifetime of a battery in hours:

(a) $P(X > 8.5) = P\left(Z > \dfrac{8.5 - 8}{\dfrac{25}{60}}\right)$

$= P(Z > 1.2)$

$= 1 - \Phi(1.2)$

$= 1 - 0.88493$

$= 0.11507$

So the probability that a battery selected at random will last more than $8\frac{1}{2}$ hours is 0.115 to 3 d.p.

(b) $P(X < 7) = P\left(Z < \dfrac{7 - 8}{\dfrac{25}{60}}\right)$

$= P(Z < -2.4)$

$= \Phi(-2.4)$

$= 1 - \Phi(2.4)$

$= 1 - 0.9918$

$= 0.0082$

The percentage of batteries that will last for less than 7 hours is 0.82%.

5 (a) Taking X to represent the number of packets in a box that contain a lucky pendant:

(i) $P(X \geqslant 6) = 1 - P(X \leqslant 5)$

Using tables of cumulative binomial probabilities with $n = 20$ and $p = 0.25$.

$P(X \geqslant 6) = 1 - 0.6172 = 0.3828$

(ii) $P(X = 6) = P(X \leqslant 6) - P(X \leqslant 5)$

$= 0.7858 - 0.6172$

$= 0.1686$

(b) Using Y to represent the number of boxes that have at least six packets containing a lucky pendant:

$P(Y = 2) = \dbinom{5}{2}(0.3828)^2(1 - 0.3828)^3$

$= 0.34452 \ldots$

$= 0.345$ to 3 d.p.

6 From calculator
(a) $r = 0.879$

(b) $y = -1.889 + 0.873x$

(c) When $x = 65$, estimated value from calculator, $\hat{y} = 55$ (2 s.f.).
Alternatively, substituting into the equation of the regression line,
$\hat{y} = -1.889 + 0.873 \times 65 = 55$ (2 s.f.)

If the assumption that physics scores depend on maths scores is correct, the estimate should be reliable, since $x = 65$ is within the range of data and the high value of r suggests a strong linear correlation between the variables.

Decision Mathematics 1

1 Applying the matrix form of Prim's algorithm gives:

	1	7	5	6	3	2	4
	A	B	C	D	E	F	G
A	–	50	40	61	–	38	–
B	50	–	–	–	–	④7	54
C	④0	–	–	35	–	–	–
D	61	–	③5	–	42	–	–
E	–	–	–	42	–	②5	10
F	③8	47	–	–	25	–	63
G	–	54	–	–	①0	63	–

The minimum cost is found by adding the circled figures and multiplying by £100.

Minimum cost = £19 500.

2

(a) The shortest distance from G to D is 16.

(b) The route is G, F, A, B, C, D.

3

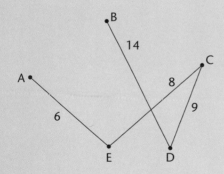

(a) The order in which the edges were connected is:
AE, EC, CD, AB (not AD or AC to avoid cycles).
The length of the minimum spanning tree is 37.

(b) Using the short-cut from B to A gives an upper bound for the TSP as 37 + 12 = 49.

4 (a)

Vertices				
Graph (i)	A	B	C	D	E
Order	4	2	4	3	3

Vertices					
Graph (ii)	A	B	C	D	E	F
Order	4	2	4	4	4	2

Vertices			
Graph (iii)	A	C	D	E
Order	3	3	3	3

(b) Graph (i) is Semi-Eulerian (2 odd vertices).
Graph (ii) is Eulerian (all vertices are even).
Graph (iii) is neither (more than two odd vertices).

5 (a) The objective function is $4a + 9b + 12c$.

(b) The formulation of the task as a linear programming problem is:
Maximise: $P = 4a + 9b + 12c$
Subject to: $6a + 15b + 20c \leqslant 100$
$a + b + c \leqslant 10$
and $a \geqslant 0, b \geqslant 0, c \geqslant 0$.

(c) In the solution a, b and c must be integers.

6 (a)

(b)

(c)

The alternating path is:
D – Wed – B – Mon – A – Tue.
The new matching is:

Notes

Index